At the Edge of Mysteries

The Discovery of the Immune System

Shantha Perera

HERO, AN IMPRINT OF LEGEND TIMES GROUP LTD
51 Gower Street
London WC1E 6HJ
United Kingdom
www.hero-press.com

This edition first published by Hero in 2022

The right of the above author to be identified as the author of this work has been asserted in accordance with the Copyright, Designs and Patents Act 1988. British Library Cataloguing in Publication Data available.

Printed and bound by CPI Group (UK) Ltd, Croydon, CR0 4YY

ISBN: 978-1-91505-452-4

Contents

At the Edge of Mysteries

For Rachel

About This Book

In December 2019 a novel coronavirus caused a devastating pandemic. Not since the Spanish-flu contagion just over a hundred years ago has a virus caused so much fear. SARS-CoV-2 threatened our lives and our livelihoods, and made us utterly vulnerable. At the time of writing, in April 2021, even though we have several effective vaccines there is no effective treatment. As new variants emerge, the efficacy of the vaccines may be reduced, threatening more waves of infection. We only have our immune system.

When I began writing this book there was no coronavirus pandemic. I wanted to write a book that enabled the general reader without a background in science to understand the immune system. My approach involved presenting the components and functions of this complex system through the lens of history, through the lives and work of the doctors and scientists who unravelled the mysteries of the immune system.

Then Covid-19, the mystery disease, appeared. I was obligated to revise the text to describe the ongoing history of the battle between the virus and the immune system. I wrote three new chapters. In the introduction, we see how the immune system responds to the virus. We return to Covid-19 at the end of the book, considering therapeutic approaches used in the past, vaccines and strategies that involve manipulating the immune system in order to tip the scales in our favour.

The book is aimed at the general reader *and* undergraduate students interested in learning about the immune system. I have included notes at the end of some chapters that go deeper into the subject matter. These paragraphs are written for the student of medical science who has some background in the subject.

Introduction

THE DOUBLE-EDGED SWORD

Beijing, January 2020. A 50-year-old man returned home after travelling to Wuhan. He had been in Wuhan for four days and was in good health, but two days after arriving back fell ill, developing a dry cough and chills. But he continued to work for another week before deciding to present himself at an outpatient fever clinic at his local district hospital. He complained of feverishness, chills, a persistent cough and shortness of breath. He was found to have a fever – an elevated body temperature of 39°C – and a chest X-ray showed 'multiple patchy shadows' in both lungs: the organs were inflamed and the alveoli, the tiny sacs where oxygen is exchanged for carbon dioxide, were full of fluid.[1] He tested positive for SARS-CoV-2, the strain of coronavirus that led to the worldwide disease of Covid-19.

He was admitted to an isolation ward, put on supplemental oxygen and given interferon alpha and two antiviral drugs. He was also prescribed an antibiotic in case he developed secondary bacterial pneumonia. As he was short of breath he was started on daily steroids to control the inflammation inside his lungs. None of these drugs were proven to be effective in treating Covid-19, but the doctors were keen to try anything to stop the progression of the disease.

Following treatment, the patient's temperature dropped to 36.4°C. But the cough, shortness of breath and fatigue did not improve. Worryingly, another X-ray taken 12 days after the onset of illness showed worsening lung infiltration. Surprisingly, his blood oxygen

7

levels were 95 per cent. (The normal range is between 95 and 100 per cent.)

Two days later he became extremely short of breath, but his oxygen levels still remained above 95 per cent. The doctors wanted to transfer him to the intensive care unit, but the man refused. He was frightened: he suffered from severe claustrophobia and did not want to be put on a ventilator. The doctors had no choice but to continue with oxygen delivered through a nasal cannula. The patient was steadily getting worse, but he kept refusing invasive ventilation.

Two weeks after his illness his shortness of breath worsened, and his oxygen saturation dropped to 60 per cent. Then he had a sudden cardiac arrest. He was immediately placed on invasive ventilation; doctors started chest compressions and gave him an adrenaline injection, but the man lost his battle with the coronavirus and died at 6.31 p.m. that evening.

Why did this 50-year-old man die? Why did he get severe pneumonia? Was it down to the virus or was the immune system responsible?

Post-mortem biopsies taken from the lung tissue reflected what his chest X-ray had indicated: damaged alveoli full of fluid containing white blood cells. He had suffered from acute respiratory distress syndrome (ARDS). Blood samples showed that the levels of two key cells of the immune system, 'helper' and 'cytotoxic' T cells, were greatly reduced even though they were activated. These white blood cells are major components of our body's defences, and a reduction in their numbers makes a person immunodeficient: unable to defend themselves against pathogens and certain cancers. But numbers of another T cell, the Th17 cell, were high. This cell, by secreting certain 'cytokines' – small proteins that cause inflammation – was in part responsible for the pathology seen in the lungs. Low levels of certain T cells and elevated levels of inflammation caused by cytokines had caused ARDS, leading to the death of the patient.

We have seen this before, in two other diseases caused by pathogenic coronaviruses: severe acute respiratory syndrome (SARS) and

Middle East respiratory syndrome (MERS). In 2003, the SARS-CoV virus infected around 8,500 people, with an overall mortality rate of nearly 10 per cent. More recently, in 2017, MERS-CoV, a more lethal coronavirus, infected nearly 2,000 people and had a frightening mortality rate of 36 per cent.[2] These three coronavirus diseases, SARS, MERS and Covid-19, caused mortality by unleashing a cytokine storm in the lungs.

*

The cytokine storm had also killed millions a hundred years ago. The last major pandemic that humanity had to endure was the Spanish flu, which claimed the lives of up to 100 million people globally. The American pathologist Jeffery Taubenberger, the man who brought the 1918 virus back to life, calls this 'arguably the single deadliest event in recorded human history'.[3] What happened then was again at the back of our minds: could it be as bad as 1918? Would there be a second wave?

This had been a different virus, H1N1, as opposed to the ones that caused SARS, MERS and Covid-19, but all these viruses have something in common. They jumped species from birds, camels and bats, posing a formidable challenge to the human immune system, which was unprepared to react to a novel pathogen. But the way some people died appeared to be very similar.

One of the mysteries of the Spanish-flu pandemic is its unusual death curve. It was W-shaped, whereby in addition to the very young and the old there was a peak in the middle. Those between 20 and 40 were highly susceptible. Young men coming home after surviving the Great War fell sick and died in their thousands. Often death came swiftly: the victims began coughing up blood and, to the horror of witnesses, their skin turned a dusky blue. Sometimes the discoloration was so profound that doctors treating infected soldiers in army encampments could not tell if they were white or black.[4]

Exactly *why* people at the peak of their lives suffered a high level of mortality continues to be debated, but what we do know is *how* they died. It was by drowning: pathologists who carried out post-mortems noted inflamed lungs, often full of bloody fluid that prevented oxygen getting into their bodies. The culprit was the immune system: a cytokine storm.

Taubenberger, working at the Armed Forces Institute of Pathology in Washington, DC, wanted to find out if there was something different about this influenza virus that had caused its unusual virulence. But first he had to find the Spanish-flu virus. He found lung-tissue samples preserved in paraffin wax taken from a 21-year-old soldier, Roscoe Vaughan, who had died in 1918. But the samples only yielded a few viral genes. He published his findings in 1997.[5]

A retired pathologist from San Francisco, Johan Hultin, came across Taubenberger's paper. In 1951, when he was a graduate student at the University of Iowa, Hultin had travelled to Brevig Mission, a small Inuit outpost in Alaska, hoping to find the 1918 virus in the bodies of villagers buried in the permafrost. Hultin hoped that the frozen conditions would have preserved lung tissues, allowing him to isolate the virus.

Brevig Mission had suffered a terrible fate as the contagion visited the tiny fishing village in the autumn of 1918. Out of the 80 Inuit villagers, 72 had perished, all within a period of five days; only eight children and teenagers survived.[6] The dead were buried in a mass grave and Hultin, after gaining permission from the village elders, started his dig. He found the bodies, took some samples and returned to Iowa City. But he failed to grow the virus from the samples.

But that was 1951. Techniques of virus gene identification had greatly improved by 1997, so Hultin was hopeful. He wrote to Taubenberger offering to return to Brevig Mission. Taubenberger agreed, and the now 72-year-old virus-hunter set off back to Alaska.

With the help of a few bemused young men, Hultin started his excavation. After finding some skeletal remains, he came across the body of a woman. The woman, whom Hultin later named Lucy, had been in her mid-twenties. She was obese and the layers of body fat had preserved her internal organs. Using a pair of garden clippers borrowed from his wife, Hultin opened her ribcage. The lungs were well preserved and full of frozen blood. He took some samples.

Afterwards Hultin filled in the grave and built two large white crosses as the original ones marking the burial had decomposed. He placed them on top of the grave and hurried back home to post the tissue samples to Taubenberger. His work was done.

Eight years later, in 2005, using lung-tissue samples from 'Lucy' and Roscoe Vaughan, Taubenberger finally isolated the eight RNA gene segments of the 1918 virus. He could now resurrect the killer that had caused the Spanish-flu pandemic.[7] Taubenberger then took the next, highly controversial step. He wanted to see if the resurrected virus was as lethal as the one that had killed up to 100 million people. So he began animal experiments.

He created two types of virus: one with all eight gene segments from the original Spanish-flu virus, and other hybrid viruses containing combinations of gene segments from the 1918 virus and those taken from a low-pathogenic influenza virus. Taubenberger infected mice, ferrets and finally macaque monkeys with these manufactured viruses. He found that there were high levels of viral multiplication in all animals exposed to the virus, but that all eight gene segments from the 1918 virus were required for maximal lethality.[8]

The macaques suffered terribly. Within 24 hours they looked depressed, refused food and drink and started coughing. They then developed ARDS and had to be euthanized eight days after infection with the virus. Post-mortems showed highly inflamed lungs, just like those of the people who perished in 1918. The monkeys had mounted immune responses, but they were insufficient to clear the virus. Instead, detrimental immune responses – the cytokine storms – had destroyed their lungs.[9]

And yet many people do survive these infections. Up to 80 per cent of those who develop Covid-19 have only mild symptoms. Their immune systems have defeated the virus. Another case will illustrate exactly how the immune system achieved this. This case also illustrates the dynamics of these protective responses.[10]

*

In early 2020, a 47-year-old woman from Wuhan, China, turned up at an emergency department in Melbourne, Australia. One week after arriving in Australia she had started showing symptoms: a sore throat, chest pain, a dry cough and lethargy. She was also feverish and had mild shortness of breath. The doctor taking her history noted that she was a healthy non-smoker who had no ailments and was not on any medication. The woman hadn't been to the Huanan seafood market, identified as a potential source of the outbreak, or had contact with any known cases of Covid-19. At the clinic, she was noted to have a fever of 38.5°C and was breathing rapidly, but her blood oxygen level was 98 per cent. However, when the examining doctor listened with the stethoscope he heard coarse, rattling sounds at the lung bases. A chest X-ray taken the next day confirmed his suspicions: there were shadows in the lungs indicating airspaces full of fluid. SARS-CoV-2 was detected by swab testing.

She was admitted and began improving over the next few days. The virus was undetectable two days after admission and a repeat chest X-ray taken four days later showed significant improvement.

Serial blood sampling revealed that her immune system was starting to fight back. Three days after admission antibody-producing B cells and anti-coronavirus antibodies appeared in her blood. Levels of these antibodies kept rising over the next two weeks. Virus-killing T cells also appeared in the blood around the same time, peaking some five days after admission. Her symptoms began resolving and she was discharged one week after she had been admitted. This patient did not require oxygen and did not go on to develop respiratory failure or ARDS. Two days later all the symptoms resolved.

Two patients, two outcomes. Both patients mounted immune responses against coronavirus: one was effective while the other was detrimental. What caused this disparity? This was one of the most pertinent questions taxing immunologists, virologists and doctors treating Covid-19 patients. And of course, all of us.

What happens when a respiratory virus infects a person? Being a parasite, the virus needs to get inside a host cell and hijack its machinery to make copies of itself. Once progeny viruses are assembled they are released from the cell and infect other nearby cells. Cycles of infection and multiplication cause a lot of cell death.

But the immune system starts to fight back. Deep inside an infected cell, the genetic material of the virus – RNA in the case of influenza and coronaviruses – is sensed by receptors. This causes activation of the cell. Signals cause proteins such as interferon regulatory factor 7 (IRF7) to trigger the expression of several genes that lead to the release of a group of cytokines called 'type I interferon'. These cytokines are involved in inhibiting viral replication and spread and generating beneficial inflammatory responses. Crucially, this first line of defence can initiate more specific immune responses: antibody and T-cell responses that clear the infection and give rise to immunological memory. Thus type I interferons have been shown to be vital in the clearance of viral infections.

But the virus is a formidable enemy. It fights back by suppressing the type I interferon response. One way it does this is by interrupting the signalling pathways used by a cell to turn on interferon genes.

The importance of type I interferons in controlling influenza was illustrated in 2015. A three-year-old girl with a mutation in the IRF7 gene, and who was consequently unable to produce enough quantities of type I interferon, presented with a severe influenza infection. She developed ARDS, had to go on a ventilator for nine days, but recovered.[11]

Experiments carried out using the reconstructed 1918 virus in macaque monkeys also showed a similar dysregulated immune response. Animals exposed to the complete 1918 virus activated fewer genes responsible for secretion of protective type I interferons such as interferon alpha (IFN-α), which led to an ineffective antiviral response and a fatal outcome.[12] SARS-CoV-2 can also do this: a SARS-CoV-2 protein called ORF3b has been shown to block the type I interferon response.[13]

Interferons are vital early in the infection. Perhaps giving the patient interferons early on might help the Covid-19 disease process? A type I interferon, interferon beta (IFN-β), has been shown to inhibit coronavirus,[14] and trials have taken place to see if could be used therapeutically. But there is a caveat to this strategy.

Once the infection has reached the lungs, possibly in part due to a suboptimal early interferon response, interferons can trigger the exaggerated inflammatory response. Interferon is a double-edged sword, so the timing of administration might be crucial.

The early 'good' interferon surge can decrease the viral load and an effective innate response can activate the big guns of the immune system. As we saw in the case of the woman who recovered from Covid-19, once antibody and T cells began to appear the virus was cleared from the system. Elimination of the virus was down to the function of antibodies and helper and cytotoxic T cells. As the levels of these antibodies and T lymphocytes increased, the patient's symptoms got better. An effective vaccine against SARS-CoV-2 must induce the generation of these antibody-producing cells and T cells. This is what happens following vaccination against the seasonal influenza virus.

But, worryingly, coronaviruses have ways of evading these specific responses as well. In a manner reminiscent of HIV, there is evidence that coronaviruses can cause T cells to commit suicide.[15] Recall that the first patient who died of Covid-19 had low lymphocyte levels. This pattern has been seen in many patients who had a severe case of the disease. High levels of pro-inflammatory cytokines also add to the depletion of T cells. Interestingly, another cytokine, interleukin 7 (IL-7), one that has been shown to increase levels of T cells, is being considered as a form of treatment to help patients with low T-cell numbers. At the time of writing, this study is ongoing.

Patients who deteriorate, typically after around seven days, may do so because of the second wave of inflammation, which is characterized by high levels of another group of cytokines: interleukins. The main interleukin in the cytokine storm is interleukin 6 (IL-6): IL-6 and other pro-inflammatory cytokines cause ARDS. The role of inflammatory cytokines in the morbidity and mortality of Covid-19 is established. Blood analysis of IL-6 levels of Covid-19 patients shows a clear pattern: those with the mildest disease had the lowest levels; those with a severe bout of the disease had more, and the patients who died had the highest.[16] The macaques in the 1918-influenza experiments also showed substantial increases of interleukin 6 when levels in the blood of infected animals were compared with pre-infection levels.[17]

*

The above account gives us a glimpse of how the immune system functions. But what are antibodies? How are they formed, and how exactly do they neutralize the virus? How do T cells work? How do they clear the infection? What role do cytokines play in the disease process? What are the organs of the immune system? What are vaccines and how do they work? What is immunological memory?

As we saw earlier, the immune system is a double-edged sword. Sometimes a response to an infection can be harmful. And in some individuals, antibodies and T cells start attacking self-tissue as if it were a transplant. Why does this happen? When and how did the immune system learn to distinguish 'self' from 'non-self', and why did that mechanism break down, leading to autoimmune diseases? Why do antibodies form against innocuous substances like grass pollen or peanuts? Can we help the immune system to destroy cancer cells?

Can we find a vaccine that neutralizes all variants causing Covid-19?

One of the best ways to understand the immune system involves revisiting the past. By going back in time, studying the key experiments and trials, we can see how this incredibly complex system was discovered. In Parts One and Two we meet the doctors and scientists and learn how their seminal discoveries unravelled the mysteries of the immune system. They were heroes and the world rewarded them with Nobel Prizes, but now, as we face yet another pandemic, we need others to step up to the mark, to come up with answers.

In the last two chapters we return to Covid-19, the disease that has paralysed the world. We see how cutting-edge discoveries in immunology are used in the ongoing battle, how vaccines were designed, and novel treatment approaches explored. And the world waits for a final victory and the end of the pandemic.

But there is another question we must ask ourselves. While we wait for a vaccine that can completely prevent infection, or effective treatment, is there anything we can do to help our immune systems to protect us from the virus? Perhaps there is. In the 1970s, scientists discovered that the immune system interacted with the brain and the endocrine system. The new field that they established, psychoneuroimmunology, showed that stress and behaviour can modulate immune responses. In the final chapter we will look at how we can translate these findings to help us in our battle with the coronavirus.

As we reach the frontiers of knowledge we are humbled by the huge gaps in our understanding of the immune system. There are many more mysteries to be solved.

PART ONE

BEGINNINGS

Chapter 1

THE SPECKLED MONSTER

Athens, 430 BC. A man sat on the cool marble steps of the Temple of Athena Nike watching the terrible scene that was unfolding in front of him. The contagion had arrived from the Port of Piraeus and was mercilessly laying waste to the defenceless city population. He sat transfixed, observing a group of people carrying the writhing body of a dying soldier. They left the man at the bottom of the steps, like some grotesque offering. His skin was red and raw, covered in small pustules and ulcers. The afflicted man rose to his feet, tore off his clothes and jumped into a drain full of dirty water to ease the pain. 'Help me!' he cried. 'Somebody help me!'

The man who rose to help was Thucydides, a general and historian. He had contracted the disease himself and had survived. He pulled the man out and laid him under a shady tree.

The Spartans had arrived in Attica, and the Peloponnesian War between Athens and Sparta, a conflict that was to last for 27 years, had gathered pace. But it was not just the ruthless Spartan war machine that was threatening Athens. A terrible plague had entered the city.

Historians still debate which pathogen was responsible for this infection, but it is possible that this was an outbreak of smallpox. What is interesting is that Thucydides, who later wrote a famous account of the Peloponnesian War, made the first known reference to immunity:

Yet it was with those who had recovered from the disease that the sick and dying found most compassion. These knew what it was from experience and had now no fear for themselves for the same man was never attacked twice, at least never fatally.[1]

Thucydides had observed a fundamental principle of immunity. Once a person had suffered and recovered from the illness they became *immunis* – Latin for 'exempt' – and gained the ability to resist a second appearance of the disease. That was why the man with the raw, red skin had been laid at his feet. Thucydides, having been infected, was now immune and would not be in danger from the illness again.[i]

We next come across a mention of immunity in the works of the ninth-century Persian physician, alchemist and philosopher Rhazes (Muhammad ibn Zakariya al-Razi), who was born in AD 854 in the city of Ray on the route of the Great Silk Road. He had studied medicine in Baghdad and was considered one of the greatest physicians of the Arab world. Students from all over the globe came to learn from Rhazes, who was greatly influenced by the Greek physicians Galen and Hippocrates, even though he had often challenged their views. Rhazes was the first physician to describe smallpox accurately, in written form, and to distinguish it from measles.

In his book, *Al-Judari wa al-Hasbah*, Rhazes writes:

The eruption of smallpox is preceded by a persistent fever, pain in the back, itching in the nose and terrors in the sleep. These are the more peculiar symptoms to approach, especially a pain in the back with fever; pricking which the patient feels all over his body; a fullness of the face which at times comes and goes; an infectious colour, and vehement redness in both cheeks; redness of both eyes, heaviness of the whole body; great uneasiness, presenting as stretching and yawning; pain in the throat and chest, with slight difficulty in breathing and cough; dryness of breath, thick spittle and hoarseness of the voice; pain and heaviness of the head; inquietude, nausea and anxiety; (with this difference that the inquietude, nausea and anxiety are more frequent in measles than in the smallpox; while the pain in the back is more peculiar to the smallpox than to the measles), heat of the whole body; an inflamed colon, and shining redness, especially an intense redness of the gums.[2]

Rhazes also wrote in his treatise that survivors of smallpox do not develop the disease for a second time.

The May 1970 *Bulletin of the World Health Organization* paid special tribute to Rhazes, stating that 'his writings on smallpox and measles show originality and accuracy, and his essay on infectious diseases was the first scientific treatise on the subject'.

The eruptions of smallpox follow a predictable pattern. The progressive rash initially appears as flat, red sores. After a few days the sores become raised bumps, which then turn into fluid-filled blisters. In the second week of infection, these pus-filled blisters crust over. Scabs form over the blisters and then fall off, around the third week of the disease, resulting in scars that are often disfiguring. Other symptoms of smallpox include fatigue, headache, body ache and occasionally vomiting. The patient can become feverish. Mouth sores and blisters can spread the virus into the throat. Complications during the illness include pneumonia and osteomyelitis (inflammation of the bone or marrow). Following illness scarring, blindness and death may occur.

Smallpox, once known as 'the speckled monster', was probably brought to Europe by the returning Crusaders from the eleventh to the thirteenth century and appeared in Britain in the sixteenth. It struck rich and poor alike: even the monarch, Queen Elizabeth I, was struck down by the disease in 1562; she survived but was left bald and with a scarred face. The seventeenth and eighteenth centuries saw deadly smallpox epidemics sweep across Britain.

In Europe some 400,000 people died annually from the speckled monster during the eighteenth century, and over a third of cases of blindness were caused by smallpox.[3] Between 20 and 60 per cent of those infected perished, and for infants the death rate was much higher: smallpox killed around 80 per cent of those infected.[4]

However, there was an age-old method that was being used in many parts of the world to combat the disease. It was called 'engrafting', or 'variolation', and involved introducing pus from a ripe smallpox pustule by means of a lancet or needle into the skin or veins of a healthy individual. This procedure was used extensively in the Ottoman Empire, and it was there that a young English aristocrat had arrived with her husband, the British ambassador, in 1717.

i Thucydides and the great physicians of antiquity had observed one of the pillars of immunity, the concept of 'specificity'. The fact that the individual was not struck down by the same disease twice, 'at least not fatally', was because the heightened immune response that conquered the infection was specific to that infection – or one caused by a very similar pathogen – only.

Chapter 2

THE GIFT OF THE OTTOMANS

Adrianople, 1717. On a warm September evening Mary made her way through the crowded Exchange, a half-mile-long street covered by an arched roof and lined on either side by more than 300 shops and cafes. Disguised in her Turkish habit, a long pair of drawers and a silk smock, over which she wore a pale-brown caftan, she walked unobtrusively, taking in the exotic sights, sounds and aromas that surrounded her. She noticed the rows of clean shops, the rich Jewish merchants haggling with their customers and the animated conversations of those who preferred to sit in the cafés drinking strong coffee or sherbet.

She arrived at her destination, a grand house belonging to a noble-woman she had befriended at the embassy. A smiling young eunuch opened the door and escorted her into the house. She was asked to wait in the large inner courtyard, where a beautiful fountain was set in the middle of a marble pond. A servant woman appeared and asked Mary to follow her. She followed the woman down a narrow passage leading from the courtyard into a room in which a crowd of women and children were gathered.

Mary Wortley Montagu was born in London in 1689. She was an English aristocrat, the eldest child of Evelyn Pierrepont, the first Duke of Kingston-upon-Hull. Mary was by all accounts a striking beauty, and in later years would become renowned for her literary skills. However, in 1713, at the age of 26, Mary caught smallpox, which left her once-beautiful skin pockmarked. She also lost her eyelashes. Her brother died of the disease in the same year and this terrible affliction would haunt her for the rest of her life. What brought her to Turkey was her marriage to Edward Wortley Montagu, who was appointed ambassador to the Ottoman Empire.

Mary Montagu described what she observed in the Turkish house in a letter written on 1 April 1717 to her friend Sarah Chiswell:

I am going to tell you a thing, that will make you wish yourself here. The small-pox, so fatal, and so general amongst us, is here entirely harmless, by the invention of engrafting, which is the term they give it. There is a set of old women, who make it their business to perform the operation, every autumn, in the month of September, when the great heat is abated. People send to one another to know if any of their family has a mind to have the small-pox; they make parties for this purpose, and when they are met (commonly 15 or 16 together) the old woman comes with a nut-shell full of the matter of the best sort of small-pox, and asks what vein you please to have opened. She immediately rips open that you offer to her, with a large needle (which gives you no more pain than a common scratch) and puts into the vein as much matter as can lie upon the head of her needle, and after that, binds up the little wound with a hollow bit of shell, and in this manner, opens four or five veins. The Grecians have commonly the superstition of opening one in the middle of the forehead, one in each arm, and one on the breast, to mark the sign of the Cross; but this has a very ill effect, all these wounds leaving little scars, and is not done by those that are not super-stitious, who choose to have them in the legs, or that part of the arm that is concealed. The children or young patients play together all the rest of the day, and are in perfect health to the eighth. Then the fever begins to seize them, and they keep their beds two days, very seldom three. They have very rarely above twenty or thirty in their faces, which never mark, and in eight days' time they are as well as before their illness. Where they are wounded, there remain running sores during the distemper, which I don't doubt is a great relief to it. Every year, thousands undergo this operation, and the French Ambassador says pleasantly, that they take the small-pox here by way of diversion, as they take the waters in other countries. There is no example of any one that has died in it, and you may believe I am well satisfied of the safety of this experiment, since I intend to try it on my dear little son. I am patriot enough to take the pains to bring this useful invention into fashion in England.[1]

This, then, was the practice of engrafting, or variolation, which was found to confer immunity to smallpox. It was widely practised in the Ottoman Empire: female Circassian slaves, who were legendary for their beauty, were much sought after by the sultans of the Ottoman Empire to populate their harems, and variolation was widely used on these girls to spare their clear, unblemished skin from the ravages of the dreaded pox. The process involved using virus that was contained in dried pus taken from a pustule of a mild case of smallpox from someone who had recovered. The variolated person often developed a mild form of the disease and was usually, though not always, immune thereafter.

Lady Mary wanted to spare her children the fate that had marked her youth and taken her brother, aged just 20. She approached the British embassy surgeon, Dr Charles Maitland, and begged him to engraft her six-year-old son, Edward. Reluctantly Maitland agreed; the boy was variolated and by all accounts he never contracted smallpox. Upon her return to London, Lady Mary tried to introduce the practice of variolation, but encountered fierce resistance from the medical establishment. In April 1721, there was an outbreak of smallpox in England and Lady Mary had her three-year-old daughter Mary inoculated by Maitland, even though the surgeon had recently retired and was living in Hertford village, some 30 miles from London.

During the peak of the epidemic, Princess Alice, the youngest daughter of the prince and princess of Wales, became unwell. At first it was thought that the princess had contracted smallpox, but this was later confirmed not to be the case. However, her illness gave Lady Mary the opportunity she had been waiting for, and together with Maitland, who was by now a convert to the procedure, she approached the royal couple. Their persistent and well-argued plea led the prince and princess of Wales to consent to what was known as the 'Royal Experiment', the first trial of variolation.

Charles Maitland was granted a Royal Licence to carry out the trial, which was overseen by Sir Hans Sloane, then president of the Royal College of Physicians as well as the court physician to King George I and Queen Anne.

In August 1721 six prisoners from Newgate prison, three males and three females, all of them condemned to death by hanging, were offered variolation. The prisoners were informed that if they survived their sentences would be commuted and they would be set free. They agreed. Maitland introduced, by incision, tiny amounts of smallpox pus into the arms and right legs of the prisoners. The following day, they each developed mild symptoms, but these disappeared in a few days and all made a full recovery. They were released. So variolation appeared to have worked, and the Royal Experiment had shown that the process was safe. But had the subjects of the trial developed immunity to smallpox?

Sloane arranged for one of the female prisoners, 19-year-old Elizabeth Harrison, who had been variolated and freed, to be deliberately exposed to smallpox. She was dispatched to Hertford village, where the epidemic was raging, and instructed to live with a ten-year-old boy suffering from active smallpox; she was to stay with him for six weeks, tending to his needs and even lying with him in the same bed. Harrison did not contract the disease, and this proved to be the first evidence that immunity to smallpox had been achieved by variolation.

Still not fully satisfied, the royal couple ordered a group of orphans to be variolated, and when these children also did not contract the disease, the prince and princess of Wales had their daughters vario-lated. This was in 1722, and from then on the practice was widely adopted by the medical profession.

Variolation soon became popular, and Sloane, evaluating the data, concluded that it did indeed protect against smallpox. The death rate from variolation was around one in 50, whereas that from the natural disease was one in six.[2]

Chapter 3

THE TEMPLE OF VACCINIA

Gloucestershire, England, 1757. An eight-year-old boy sat huddled together with five others in a locked barn. For the past six weeks they had been purged, bled and fed a diet low in vegetables: they were being 'prepared'. A smallpox outbreak had prompted many people to have their children variolated. The boy's name was Edward Jenner.

The process made young Jenner quite ill, and he emerged from the barn looking like a ghost: pale, weak and thin. The 'preparation', it was believed, was crucial for removing impurities before variolation. The procedure, as it was practised in those days, was expensive and few could afford it. It was also not without risk: some became blind or deaf, or suffered from terrible arthritic pains. Jenner survived. But the experience had left an indelible scar on his psyche: a fear of smallpox.

Edward Jenner was born in Berkeley, Gloucestershire, in 1749. His father was the vicar of the town, and young Edward received a good education. But at the age of five he was orphaned and went to live with his older brother. In eighteenth-century England smallpox was rife and was the biggest killer of adults. Most victims suffered a flu-like illness, developed a rash that turned into blisters and then succumbed to the disease. Those who survived suffered terrible complications, most notably deafness and blindness.

Lady Mary Wortley Montagu died in 1762. Not long after, in 1770, Jenner, now aged 14, embarked on a career in medicine. He was apprenticed for seven years to surgeon Daniel Ludlow. While observing a consultation with a female patient at Ludlow's surgery, Jenner heard her say, 'I shall never have smallpox for I have had the cowpox. I shall never have an ugly, pockmarked face.'[1]

At the age of 21, Jenner entered the next phase of his medical education and became a pupil of the famous surgeon and researcher John Hunter. Jenner lived at Hunter's house and attended St George's Hospital in London. He was greatly influenced by Hunter, who, in addition to being a great anatomist and inveterate collector of pathological specimens, was a keen naturalist. Jenner himself had always been interested in ornithology and, encouraged by Hunter, carried out some interesting studies on the breeding habits of cuckoos, research that would later lead to a prestigious fellowship of the Royal Society.

After qualifying as a surgeon-apothecary in 1733, Edward Jenner returned to Berkeley and established a country practice. Smallpox was still a significant problem, with 60 per cent of the population contracting the disease and 15–20 per cent succumbing to it.[2] Patients often asked Jenner to variolate them or their children against the pox, and the doctor complied. Variolation had been improved since his own childhood experience; Jenner followed the method introduced by Dr Robert Sutton, who carried out the procedure by injecting a small amount of pus taken from a smallpox pustule into the skin. This was a relatively painless method and did not usually draw blood.

Jenner's practice regularly brought him into contact with farm workers. He observed that, as Ludlow's patient had said, milk-maids and milkmen did not appear to contract smallpox if they had previously become infected with cowpox, although this did not apply in every case. Jenner also noted that only some types of cowpox conferred resistance – he called this 'true cowpox'. The thought of using cowpox material for variolation, which would be a lot safer, crossed his mind and he consulted his former mentor, Hunter, for advice. Hunter's response to his young former pupil characterized his approach to all such enquiries. 'Interesting observation, Jenner,' Hunter said. 'But why think; why not do the experiment?'[3]

On 14 May 1796, Edward Jenner did just that. After obtaining parental permission, he inoculated an eight-year-old boy, James Phipps, with pus taken from a cowpox pustule on the arm of Sarah Nelmes, a milkmaid who had contracted the disease from her cow, Blossom. Phipps subsequently had a low-grade fever but recovered in a few days, and on 1 July 1796 Edward Jenner inoculated the boy with live smallpox. But James did not contract smallpox. In fact, over the next few years young Phipps would be variolated some 20 times and demonstrated a persistent immunity to the disease.

Jenner repeated this experiment on 23 other patients, and in 1798 published his findings.[4] He paid for the publication out of his own pocket. This was the first demonstration that cowpox, a similar but different microorganism, had conferred immunity to smallpox. He decided to call the method 'vaccination', derived from the Latin word for cow, *vacca*. There was initial resistance from the medical establishment, however. Variolation, as it was practised at the time, was financially lucrative, although many physicians eventually began carrying out the new procedure. But many patients were fearful of the effects of cowpox, some even believing that using this material would give them bovine features. Jenner was parodied in cartoons that showed a throng of vaccinated people surrounding the doctor, many of them sprouting horns and udders.

But Edward Jenner had friends in the aristocracy, and the fifth Earl of Berkeley used his influence to help him get vaccination off the ground. Once its efficacy and safety were appreciated, the practice began to spread throughout the country, and in 1853 Jenner's vaccination was made compulsory in England. Jenner, however, did not seek to enrich himself and continued to practise at Berkeley, where his summer house, now named the 'Temple of Vaccinia', was used to vaccinate the poor without charge.

As vaccination was adopted throughout Europe and America, Jenner's fame soon spread. When England went to war with France, Jenner was asked to write to Napoleon to secure the release of some English prisoners. The emperor agreed, stating that he 'could not refuse a request from such a great benefactor of mankind'.[5] But around the rest of the world there were many who could not afford vaccination and, even in the 1950s, some 50 million cases of smallpox appeared annually.[6]

Edward Jenner died in Berkeley in 1823, aged 74. He had discovered a relatively safe and effective method that could protect against the speckled monster that killed millions. He had given the world immunization. Eventually, Jennerian vaccination spread throughout the word and the World Health Organization, following a global eradication programme, declared the eradication of smallpox in 1979.[i] Interestingly the vaccine used was essentially the same as that used by Jenner. His house in Berkeley is now a museum and visitors can wander through his study, walk in the manicured garden and gaze at the odd-looking summer house where he vaccinated the poor for free.

---

i   Vaccines have had a profound impact in reducing the incidence of several diseases. These include diphtheria, measles, mumps, pertussis (whooping cough), rubella, poliomyelitis, tetanus and smallpox. What gives protection in these vaccines is a heightened response generated by a population of cells called 'memory cells'. The nature of the memory-cell response is discussed in Chapter 24.

Chapter 4

THE PREPARED MIND

The wolf entered the village from the mountains of Jura. The boy saw the creature with his own eyes as it ran through the village biting dogs, chickens and even the men who desperately tried to kill it. The wolf was mad, driven insane by rabies, the destroyer of dogs and men. Finally, they killed it and burned its body.

But the dogs that had been bitten then also began to show signs of the disease. First their barks changed in tone; they developed fever, lost their appetite. Then some entered the mad phase, viciously savaging any animal or person that crossed their path. Some of these dogs were caught and put into cages, but they immediately began chewing the wire. They appeared to have an insatiable appetite and were constantly barking. These dogs feared no one. But then that phase passed, and dejection set in: their jaws dropped, they began to drool and foam around the mouth, and finally they died.

The boy remembered a school friend who was bitten by one of these dogs. He was without symptoms for around a month, but then a profound exhaustion set in and he developed a fever. After that he became highly anxious. After a week he suddenly became confused, hyperactive, agitated. He greatly feared water and began salivating. Then he suffered convulsions. There were, however, periods during which he regained his orientation and realized what was happening to him. Mercifully he slipped into a coma and died a few days later.

The young boy who witnessed these events in the Jura would later become one of the greatest scientists of all time. Edward

Jenner's observational studies on smallpox had given the world vaccination against the disease. But no one knew of the existence of the disease-causing microorganisms, now called pathogens, which were the cause of many devastating diseases in animals and humans.

This young Frenchman would take the research much further, creating vaccines for several of the most important diseases that affect both livestock and humans. He would also carry out the first controlled clinical trial of a vaccine, and justifiably come to be called the father of the new science of immunology. He was also in part responsible for formulating the germ theory of disease. For the first time the enemies of mankind, which laid waste to millions every year, were shown to be an army of living, reproducing microorganisms whose lethal grip could be tempered by harnessing the power of the immune system. His name was Louis Pasteur.

*

Pasteur was born in 1822 in Dole, France, just one year before the death of Edward Jenner. His father, an ardent nationalist, was a sergeant major in Napoleon's army, and young Pasteur was from an early age imbued with the glory and majesty of the French Empire. The family moved to Arbois in 1827, where Pasteur was enrolled in the local school. After completing his secondary education, Pasteur entered the Royal College of Besançon, where he obtained his Bachelor of Arts and Bachelor of Science degrees. Further studies led to a doctorate from the École Normale Supérieure in Paris in 1847.

Pasteur trained as a chemist, and, in 1854, aged 32, became a professor and dean of the Faculty of Science at Lille University. In 1856, he was asked to solve a problem at a local distillery. The beer was going sour and the owners had no idea why. The production of beer involves adding yeast into a solution of sugar beet, and at the time the ensuing fermentation was thought to be a purely chemical process despite a living viable organism – yeast – being present in the brew. Pasteur believed that fermentation was in fact biological and that the living yeast somehow caused fermentation to occur. The souring, he supposed, was caused by another microorganism that had somehow got into the vats. He found rod-shaped microorganisms in the scum that had formed at the edges of the vats. He isolated and grew them, observing how they multiplied. After careful experimentation, Pasteur found the cause of the souring: lactic-acid bacteria. He was intrigued by these minute organisms that were found in the spoiled beer. But he wasn't the first person to observe microscopic organisms: Antonie van Leeuwenhoek had first described them back in 1677 using his hand-crafted microscopes. He had called them 'animalcules': tiny animals.[1]

But where did they come from? At the time the prevailing view was that these germs were spontaneously synthesized from decaying matter: the theory of spontaneous generation. Pasteur was determined to demolish this theory.

In October 1857 Pasteur was appointed director at the École Normale Supérieure and, along with his wife Marie and three children, left Lille and moved to Paris.

In 1859, the year of the publication of Darwin's revolutionary *On the Origin of Species*, Pasteur's first-born daughter Jeanne contracted typhoid. At that time, when the average life expectancy at birth in Europe was just 40, typhoid was rampant and often deadly. The disease now claimed Pasteur's daughter, and nine-year-old Jeanne died in September that year.

Pasteur threw himself into his work. He was convinced that microbes in the air could contaminate any liquid they came into contact with – for example, milk, wine or vinegar. But the medical establishment, still entrenched in the idea of spontaneous generation, disagreed, and Pasteur was ridiculed, often in public.

Undeterred, he carried out a series of ingenious experiments using swan-necked flasks to show conclusively that the microorganisms from the air could contaminate sterile liquids. In one such experiment Pasteur boiled and thus sterilized an infusion of broth in a swan-necked flask. Even though the flask was exposed to the air the infusion remained sterile: dust particles containing microorganisms from the air were trapped in the swan-neck curve of the flask and could not enter the broth (Fig. 1).

In April 1864, Pasteur described this experiment to a spellbound audience at the Sorbonne:

> By boiling, I destroyed any germs contained in the liquid or against the glass; but that infusion being again in contact with air it becomes altered, as all infusions do. Now suppose I repeat this experiment, but that, before boiling the liquid, I draw the neck of the flask into a point, leaving, however, its extremity open. This being done, I boil the liquid in the flask, and leave it to cool. Now the liquid of this second flask will remain pure not only two days, a month, a year, but three or four years – for the experiment I am telling you is already four years old, and the liquid remains as limpid as distilled water.[2]

When he tipped the flask, allowing the infusion to enter the swan neck, the broth became infected, proving that the microorganisms, rather than being generated spontaneously in the broth, were in fact trapped in dust particles carried by the air that now contaminated the broth. These famous experiments sounded the death knell for the theory of spontaneous generation and ushered in a new idea, one that Pasteur called the germ theory of disease.[3]

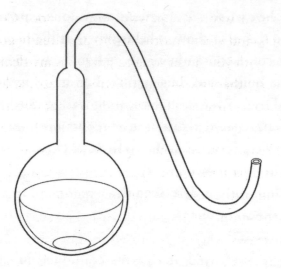

Figure 1. Swan-necked flask.

Pasteur also showed that the microorganisms that caused spoilage could be destroyed by heating the liquids.[4] The age of 'pasteurization' had arrived.

In April 1864, Pasteur felt that he had all the data he needed to prove his germ theory, and he presented his findings to a group of influential scientists at the University of Paris. It was a climactic moment and Pasteur's presentation was flawless. It was difficult for the establishment to argue against such compelling scientific data, and Pasteur's theory was accepted. His fame spread throughout Europe.

But, once again, tragedy struck the Pasteur household. In June 1865, Pasteur's father, Jean-Joseph Pasteur, the old soldier of Napoleon, died. Then, in September 1865, his daughter Camille died of a liver tumour. Pasteur, it was said by one of his biographers, 'was stunned by pain'.[5] But this was not to be Pasteur's final visit to the small cemetery at Arbois. In May 1866, his third daughter, Cécile, was also taken by typhoid fever, aged just 12.

'For the first time in his life', writes his biographer Patrice Debré, 'Pasteur gave in to despair.' 'So, they will all die one by one, our dear children,' Pasteur wrote to his wife, 'you my poor Cecile whom I loved so much, and you two others who are already gone and who call her to be with you. I too long to join you, my dear children.'[6]

Pasteur's germ theory was generally recognized, at least as far as brewing was concerned, but few considered that infectious diseases were also caused by microbes. It was still widely believed that such diseases were caused by miasma, or bad air. There were no effective treatments, and patients were often offered little more than leeches and rooms full of the fumes of burning sulphur.

Extending his germ theory into disease, Pasteur now argued that infectious diseases were also caused by microbes that were carried in the air. Each disease, he believed, was caused by a unique microorganism.

In 1868, weighed down by grief and a relentless schedule, Louis Pasteur suffered a stroke that left him partially paralysed and forced him to cut back on his work. He too was offered leeches. Even so, in 1869 Pasteur was asked to find the cause of a mysterious disease that was killing silkworms in Cévennes. He discovered the causative agent, a microbe, by transferring the disease through injection of microbes isolated from diseased worms into healthy worms. This was the first time that an animal disease was experimentally proven to be caused by a microorganism.

In 1873, Pasteur was appointed to the French National Academy of Medicine. But it was not Louis Pasteur who would prove beyond doubt that a human disease was caused by a microbial pathogen. That discovery came from work done in Germany by a country doctor called Robert Koch.

i   The Italian physician Girolamo Fracastoro had suggested in 1546 that some diseases were caused by minute bodies that multiplied.

Chapter 5

THE COUNTRY DOCTOR

He worked alone. That's what he preferred, to work in the quiet laboratory late into the night, undisturbed. He rarely told anyone, even his close colleagues, what he was doing, which experiments he was performing. Methodically recording all his findings in his notebook, he continued, in his private world. That's how he liked it. His wife had long grown accustomed to him coming home in the small hours. An icy formality had cloaked their marriage. She appeared to have lost interest in him. He was relieved as the thought crossed his mind. No matter. There was Hedwig. There was always Hedwig. He had found it difficult to get the 19-year-old who had come to photograph the institute, and who had spoken to him, out of his mind. There appeared to be a spark of interest in her grey-green eyes... or was it all his imagination?

Robert Koch was born in 1843 in Clausthal, northern Germany. Highly intelligent, young Koch taught himself to read at the age of five; he told his bemused parents that he had achieved this feat by reading local newspapers. He received his doctorate in medicine from the University of Göttingen in 1866. There, Koch was influenced by Jakob Henle, professor of anatomy, who in 1840 had suggested that infectious diseases were caused by living organisms. After graduation Koch served as a surgeon in the Franco-Prussian War, which took place between 1870 and 1871. After the brief conflict, which saw Germany emerge victorious, Koch decided to become a general practitioner and settled, with his wife Emma and daughter Gertrude, in Wollstein, Posen (now Wolsztyn in

Poland). By all accounts Koch was a good family physician and his practice thrived.

But Koch found the work dull and uninspiring, and so decided, in his spare time, to study anthrax, a disease that was endemic in the region, and that killed tens of thousands of livestock every year.[1] Infected cattle and sheep would suddenly develop a fever along with a harsh cough that was often bloody. The animals, Koch noted, looked wretched, with staring eyes, shivering and suffering fits. They would lose their appetites, and unsurprisingly milk production would cease. Death would soon follow.

Humans too could become infected with anthrax, around a hundred dying from the disease every year. Most contracted the disease through the skin, the infection manifesting as an unsightly dark sore. More rarely, anthrax was acquired through ingestion of meat from infected carcasses, and this form of the disease was generally fatal. However, by far the deadliest form of anthrax was the one transmitted by inhalation. Anthrax spores, once inhaled, spread throughout the body. Initial symptoms included a bloody cough, shortness of breath, fever and chest-wall pain. Invasion of the lymphatic system – the extensive network of thin-walled vessels containing immune cells that traverses the body – caused the spores to spread rapidly, with swelling of lymph glands commonly seen. But it was the toxins produced by these spores that caused death. We will come across these lymph glands again in Chapter 23.

Koch began carrying out research at his home, in a small, makeshift laboratory he had set up, separated by a curtain, in a corner of his consultation room. He had read that there were rod-shaped microorganisms in the blood of sheep that had died of anthrax, and, using a simple microscope – a birthday gift from Emma – he tried to find out if these microorganisms were the causative agents of anthrax.

He injected a mouse, one of Gertrude's unfortunate pets, with blood taken from a sheep that had died of anthrax. Dissecting the mouse and examining its tissues, Koch found rod-shaped microbes in the blood, spleen and lymph nodes. He then took some bacteria-laden blood from the mouse and inoculated another healthy mouse. This second mouse also contracted anthrax, and Koch was able to find anthrax organisms in its tissues. He repeated this many times and managed to transfer the disease some 20 times, always finding the strange, twisting pathogens in the blood of the infected animals. But the blood also contained other organisms, and Koch needed to prove that it was only the rods that were responsible for anthrax. He needed to find a way to separate them from the other contaminants.

His ingenious method of isolation involved placing a drop of the sterile clear fluid – the aqueous humour – drawn from the eye of a freshly killed ox onto a hollow that had been ground into a glass slide. He inoculated this with a suspension of macerated spleen tissue from an infected mouse. Warming the slide to 40°C, Koch found that within a few hours the bacteria from the spleen had begun to thicken and elongate, forming long filaments that eventually transformed into spherical spores. These spores, Koch observed, remained in the ox fluid long after the filamentous forms had disintegrated. If the culture was dried and reconstituted with fresh aqueous humour, the rods reappeared. They had germinated from the spores! This finding also explained the presence of dormant anthrax in soil and how certain areas of land remained toxic for many years. Koch photographed the anthrax microbes he had managed to isolate in pure culture. By injecting healthy animals with pure cultures of anthrax bacilli, Koch was finally able to prove beyond doubt that the bacilli caused anthrax. As had first been suggested by Pasteur, Koch was able to prove that a disease was caused by a specific microbe.[2]

The general practitioner and part-time researcher from Wollstein gathered up his slides and notes and headed for Breslau, where he demonstrated his techniques to Ferdinand Cohn, then professor of botany at the city's university. Realizing that he was witnessing something truly remarkable, Cohn summoned his colleagues – among them Julius Cohnheim, professor of pathology – to witness Koch's demonstration. Encouraged by Cohn and Cohnheim, Koch published his findings in 1876.[3]

Four years later, the significance of his work now fully recognized, Robert Koch left his practice at Wollstein and took up a position at the Imperial Hygienic Institute in Berlin.

Chapter 6

THE CHALLENGE

In France, Louis Pasteur came across Edward Jenner's publication. As he read the paper, Pasteur realized where his destiny lay: he would develop vaccines against infectious diseases. Now that the microorganisms responsible for many infectious diseases were being isolated, this ambition had become a real possibility.

The development of vaccines in many ways first required the establishment of the germ theory of infectious disease. Jenner had used pus taken from a cowpox blister to prepare his vaccine, but most infectious diseases did not cause blisters and thus the pathogens could not be readily isolated and used in a vaccine. Pasteur began by studying the cowpox virus and attempted to isolate the microbe. But the fact that this was a virus – unknown to Pasteur – meant that it could not be isolated by conventional methods used at the time to culture bacteria.

Pasteur then turned his attention to another infection that was causing problems to the farming industry at the time: chicken cholera. The disease had decimated the fowl population in Europe, and was also causing disease in turkeys and waterfowl. The diseased birds had ruffled feathers and suffered from chronic diarrhoea and swollen joints. They lost their appetites and died suddenly from an overwhelming blood infection. Pasteur had isolated and grown the causative agent, thereafter named *Pasteurella multocida*, in 1880.[1] Now he attempted to make a vaccine against chicken cholera.

In an early experiment, one of Pasteur's assistants, a young physician called Charles Chamberland, was intending to inoculate a batch of chickens with a live cholera vaccine. Rushing to finish all his tasks before going on holiday, Chamberland forgot to inoculate the birds. On his return, noticing the unvaccinated chickens, he quickly inoculated them using the remains of the vaccine preparation that had been left on the bench for two weeks. Then, as instructed by Pasteur, he injected all the chickens with live cholera. The chickens survived! Chamberland admitted his transgression to Pasteur but, to the young assistant's surprise and relief, Pasteur didn't reprimand him. Pasteur was more interested in why the chickens had survived. He asked for the experiment to be repeated, but this time he wanted the vaccine culture to be deliberately left on the bench for two weeks before inoculating a fresh batch of hens. Chamberland repeated the experiment. All the birds survived the cholera injections. 'We may have a vaccine, Chamberland,' Pasteur remarked. 'Perhaps I should send you on holiday more often!'

This was a truly original finding. For the very first time a vaccine had been made using a microorganism cultured outside the host. It was also the first time a vaccine had been used against a disease whose causative agent had been isolated and identified. Chamberland's oversight had been seized upon by Pasteur and the process of attenuation had been discovered. 'Chance,' as Pasteur once told his students in a lecture at Lille, 'favours the prepared mind.'

As Rhazes and Thucydides had observed, and as the Royal Experiment and Jenner's and Pasteur's vaccinations had demonstrated, a heightened and more effective immune response follows a second encounter with a pathogen (Fig. 2). This is called the secondary response; the initial encounter with the pathogen either by natural infection or vaccination provides the primary response.

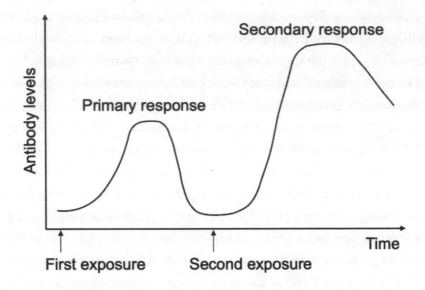

Figure 2. Primary and secondary immune responses.

The generation of a protective secondary response is the goal of the vaccine strategy. But, as we saw with Jenner's vaccine, the vaccine strain doesn't necessarily have to be the same microorganism that causes the disease. A related organism can give rise to this effective and heightened secondary response. Alternatively, a mild form of the pathogen, as in the case of variolation, or a deliberately weakened form of the pathogen, as in Pasteur's cholera vaccine, can also be as effective. However, the details of why or how this happened and the cells and molecules that were involved in the secondary response remained a mystery for decades to come.

Pasteur's next challenge was anthrax. Robert Koch had isolated and characterized the pathogen, but anthrax was a major problem for agriculture in France at the time. Just as he had done with cholera before, Pasteur began experimenting on methods of attenuation, attempting to weaken the pathogen to be used as a safe but effective vaccine. Simply ageing anthrax cultures, he found, did not attenuate the pathogen: aged cultures of anthrax produced spores and were still dangerous. Spores had to be heated above 42°C to be killed. By conducting a series of experiments, Pasteur found that if anthrax cultures were kept between 42 and 43°C for a week, they failed to sporulate and were therefore safe to use as a vaccine.[2]

Pasteur successfully vaccinated guinea pigs, rabbits and sheep using this method of attenuation. After these studies were complete, he was confident he had a vaccine against anthrax. In honour of Jenner, whom Pasteur considered his predecessor, the anthrax culture he used was called a 'vaccine'.

The response of the veterinary establishment, however, was luke-warm. Pasteur's germ theory still had vociferous opponents who questioned its relevance to animal and human disease. Hippolyte Rossignol, the influential editor of the *Veterinary Press*, wrote scathingly about Pasteur's assertion that diseases were caused by microbes, and the president of the Agricultural Society of Melun, Baron de La Rochette, issued a challenge to Pasteur. 'Prove it. Prove that your vaccine will work against anthrax,' he said. 'You can use my farm to do your experiment.'[3]

Pasteur accepted the challenge. He believed that he had found an effective vaccine against anthrax and was ready for a field test. The stage was set for the first controlled clinical trial in immunology, which would put Jenner's vaccination hypothesis to the test.

The trial began on 5 May 1881, on Baron de La Rochette's farm at Pouilly-le-Fort. Twenty-four sheep, a goat and six cows were injected with an attenuated culture of bacteria. Twenty-five sheep were not inoculated as a control. It is interesting to note that the method of attenuation was not specifically mentioned in Pasteur's original paper, published later that year. Pasteur's laboratory books revealed years later that the method used was a chemical attenuation using potassium bichromate, a method devised by a competitor called Jean Toussaint. Pasteur had not used the temperature method or oxygen attenuation, techniques he had successfully used in his preliminary experiments. The experimental animals were given a second inoculation of a stronger vaccine culture on 17 May.

The efficacy of the vaccine was tested on 31 May, when Pasteur's assistant completed the experiment by injecting a fully virulent anthrax culture into both the vaccinated and the unvaccinated animals.

That evening several animals, including some of the vaccinated ones, became feverish. For a moment it looked as if the experiment had failed. But a telegram arrived on 2 June to inform Louis Pasteur that the vaccinated animals had recovered and that the experiment appeared to be a success. Pasteur left for Pouilly-le-Fort early the next day. A crowd had gathered at the farm when Pasteur arrived. In his paper, Pasteur describes what happened next:

> When the visitors arrived on June 2, they were astounded. The twenty-four sheep, the goat and the six cows which had received the vaccinations of attenuated anthrax all appeared healthy. In contrast, twenty-one sheep and the goat which had not been vaccinated had already died of anthrax; two other unvaccinated sheep died in front of the viewers, and the one remaining sheep died at the end of the day.[4]

The crowd burst into applause. Pasteur was victorious. His vaccine had worked. This, then, was the first demonstration, by a controlled experiment, of active immunity achieved by vaccination. Louis Pasteur had validated Edward Jenner's hypothesis. In his publication, Pasteur writes:

> Each one of our attenuated anthrax microbes is a vaccine for the virulent microbe, that is to say, an adapted virus which produces a more benign variety of disease...
> We now possess a vaccine of anthrax which is capable of saving animals from this fatal disease; a virus vaccine that is itself never lethal; a live vaccine, one that can be cultivated at will and transported without alteration. Finally, this vaccine is prepared by a

procedure that we believe can be generalized since, the first time around, this was the method we used to develop a fowl cholera vaccine. Based on all the conditions that I list here, and by looking at everything only from a scientific point of view, the development of a vaccination against anthrax constitutes significant progress beyond the first vaccine developed by Jenner, since the latter had never been obtained experimentally.[5]

Following the trial at Pouilly-le-Fort, Pasteur's live attenuated vaccines were adopted with enthusiasm and greatly reduced the mortality of sheep and other farm animals from anthrax.

*

In Berlin, Robert Koch read Pasteur's publication, his mood darkening with every line. He simply could not bring himself to believe it. An unapologetic nationalist, Koch despised the French. The feeling in Paris was mutual. The Franco-Prussian War was over, but the battle of medical science had only just begun.

Koch had developed, along with Walther and Fanny Hesse and Julius Petri, a novel medium for culturing and isolating bacteria. A solid, jelly-like medium, agar, made it possible for Koch to obtain pure cultures of microbes with greater ease, a feat that greatly facilitated the identification of causative agents of many diseases. A humble, circular dish (named a 'Petri dish' after its inventor, and still widely used today) had arrived and would play a vital part in the ongoing research.

In the summer of 1881, Louis Pasteur and Robert Koch met for the first time in London, where Joseph Lister, the pioneer of antiseptic surgery, had invited them to the Seventh International Medical Congress. Despite their earlier antipathy, the meeting was cordial. Koch dazzled the audience with demonstrations of his staining and plate culture techniques. Pasteur was impressed: '*C'est un grand progrès, Monsieur*,' he told the German as he finished his lecture.[6] After Koch returned to his seat Pasteur took to the stage himself. He proceeded to present his data on the trial conducted in Pouilly-le-Fort. Now it was Koch's turn to be impressed: the data was unambiguous. The vaccine had worked.

But the warm cordiality between the French chemist and the German doctor that summer in London did not last and the two giants of bacteriology and immunology spent much of the rest of their careers attacking each other. Koch was critical of Pasteur's methods. 'He is using impure cultures,' he remarked at a conference, adding contemptuously, 'Pasteur is not a physician... one cannot expect him to make sound judgements about pathological processes and symptoms of disease.'[7]

Pasteur in turn barely concealed his contempt for German science and for Koch in particular. In 1868 Pasteur had been awarded the honorary degree of Doctor of Medicine by the University of Bonn. But three years later, incensed by the German invasion and occupation of Paris, he returned the degree along with a vitriolic letter asking the university to 'efface my name from the archives of your faculty and to take back that diploma, as a sign of the indignation inspired in a French scientist by the barbarity and hypocrisy of him who to satisfy his criminal pride persists in the massacre of two great nations'.[8]

The Germans retaliated and the war of words continued unabated. Nevertheless, Pasteur's method of anthrax attenuation was eventually adopted in Germany. The French school led by Pasteur focused on vaccines, while the German school under Koch was more concerned with isolating pathogens and bringing forth public-health measures to prevent disease.

Let's look at Pasteur's approach to preparing vaccines, which involved the principle of attenuation.[i] Pathogens can be attenuated using a variety of methods. In the case of viruses, serially passaging the microorganism through a series of isolated tissue cultures or growing them in chick embryos can make the virus progressively more adapted to growing in these tissues and therefore less able to grow and replicate as effectively inside the human host. These second-generation viruses will therefore be less pathogenic in humans. However, they will still be as immunogenic as the original pathogenic strain: that is, they will be able to provoke a vigorous immune response, which is the essential requirement of a vaccine. This is because the antigens – substances that when introduced into the body of an animal induce the production of an antibody – that induce the immune response remain intact following attenuation. But there is a drawback. These 'live' vaccine viruses can, by mutating, potentially revert and become pathogenic once more. The oral polio vaccine that used to be given to children was made from live polio virus that was attenuated. However, because mutations can occur commonly in the polio virus genome, rare cases of paralytic polio occurred because of using the live vaccine. Currently this vaccine has been replaced by a killed or inactivated polio virus to avoid this potential danger of reversion.

So why not simply use inactivated or killed viruses as vaccines? You could easily kill the pathogen by using heat or a chemical such as formaldehyde. This would clearly be a safer strategy, but again there are disadvantages. Heat and chemical treatment can disrupt the antigens, thereby making the strain less immunogenic. Sometimes it is important that the vaccine strain grow and replicate inside host cells in order to stimulate the cells of the immune system.[ii]

But some vaccines in use today are made from killed or inactivated microorganisms. An example of a vaccine that uses inactivated viruses is the trivalent influenza vaccine, which is made up of antigens from three common influenza strains. However, another influenza vaccine currently in use and administered as a nasal spray is an attenuated live viral vaccine. It is also possible to use a related microorganism as a vaccine. Jennerian vaccination used cowpox virus, and its effectiveness was because of the shared antigens between smallpox and cowpox viruses (Fig. 3).

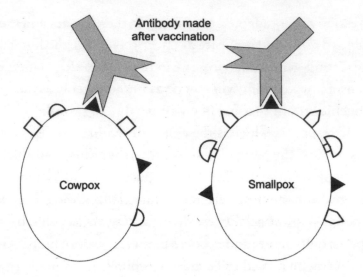

Figure 3. Smallpox and cowpox: shared antigens
and protective antibody.

i   Attenuation involves a process whereby the virulent pathogen is grown under abnormal culture conditions or in different host tissue to select for avirulent organisms. An example of such an approach is the BCG strain of tuberculosis (TB) grown in increasing bile concentrations. Another example of an attenuated live vaccine is the measles, mumps and rubella (MMR) vaccine.

ii   Live viral vaccines replicate inside host cells and present viral antigens to the host T cells to generate T-cell responses that enable the destruction of the infected cell. Live vaccines provide prolonged exposure of pathogen epitopes to immune cells, which increases immunogenicity and memory-cell production. Problems of using live vaccines include reversion and contamination with other viruses derived from the tissue used in its preparation that may harm the host. Inactivated viruses cannot do this as they cannot infect cells, get inside them and display their antigens to T cells. So, although these vaccines can elicit good antibody responses that can block the virus attaching to host cells, they are less able to get rid of infected cells. Also, since they cannot replicate, vaccines that use killed viruses will need to be given repeatedly, hence the need for boosters.

Chapter 7

THE BITE OF DEATH

The boy who had witnessed the carnage left by the rabid wolf in the Jura was now a 59-year-old man who walked with a limp. But Louis Pasteur, somewhat recovered from his stroke, now began his last great battle: to find a vaccine against rabies.

Pasteur believed that the rabies microbe passed from dog to man through infected saliva. But as it was a virus, he could not isolate and identify the microorganism. Once again, however, serendipity played its part. Completely by chance, Pasteur had come across some experiments carried out by one of his most gifted assistants, Dr Émile Roux, who had been part of the team that had carried out the trial at Pouilly-le-Fort.

Roux was born in Confolens in south-western France in 1853. After completing his medical studies, he went to work for Louis Pasteur in 1878. He was given the task of developing a vaccine against rabies. First, Roux needed a plentiful supply of virus, and so he set about investigating various methods to concentrate and harvest the microbe. He had found that spinal-cord suspensions prepared from rabies-infected animals were teeming with the virus, and when macerated and injected into healthy animals reproduced the disease. The spinal cords were stored in liquid in jars, and it was these jars containing concentrated rabies virus that gave Pasteur an idea.

Without bothering to inform Roux, Pasteur began 'passaging' the virus: injecting spinal-cord tissue from an infected rabbit into the brain of a healthy rabbit, and after that rabbit developed rabies repeating the procedure in a fresh animal. Pasteur found that this method of serial passaging of the virus from one animal to another led to increased virulence of the virus in rabbits but decreased virulence in dogs. Drying the spinal cord was found to further attenuate the virus.

He took an infected rabbit spinal cord that had been dried for 14 days, liquefied it and injected the preparation into a dog. The dog did not contract rabies, even after it was given progressively stronger doses. From this study, Pasteur devised a vaccination protocol. After drying for two weeks a rabbit spinal-cord preparation, found to be non-infectious in dogs, was used as the first dose of a vaccine. The dogs were then injected with further doses of cord that had been progressively dried for shorter periods and thus made increasingly virulent. The last dose administered contained fully virulent spinal cord.

Using this method, Pasteur created a rabies vaccination protocol that protected dogs from developing the disease even after being bitten by a rabid animal.[1] By 1884, he had announced the arrival of an efficacious rabies vaccine for dogs. The long incubation period of rabies meant the vaccine could be administered even after the animal had been bitten. But Pasteur was hesitant to try out his vaccine on a human subject. Perhaps, he mused, he should try it on himself first.

But in July 1885 a woman burst into his laboratory with her son, who had been recently bitten by a rabid dog. Even though he wasn't a medical doctor, Pasteur realized that the boy's condition was serious. He counted 14 separate bite wounds, and there was no question of the prognosis: he would surely die of rabies. The mother pleaded with Pasteur to treat him with the experimental vaccine. News of the success of the dog vaccine had spread throughout Europe. But Pasteur refused. 'The vaccine has never been tested in a human,' he said. 'It's a vaccine for dogs, Madame, not humans.'

The desperate mother continued to plead with him. 'Please, Monsieur Pasteur!' she cried. 'You are the boy's only hope.' Finally, her pleas wore him down and, with great reluctance, he agreed. Not being a physician, Pasteur could not carry out the injection himself, and so he asked an assistant who had a medical degree to administer the vaccine to the boy. Joseph Meister, aged nine, thus became the first person in the world to receive Pasteur's rabies vaccine: an injection of dried rabbit spinal cord, liquefied and suspended in veal broth. Pasteur describes this climactic moment in his own words:

> The death of this child appearing to be inevitable, I decided, not without lively and sore anxiety, as may well be believed, to try upon Joseph Meister the method which I had found constantly successful with dogs. Consequently, 60 hours after the bites, and in the presence of doctors Vulpian and Grancher, young Meister was inoculated under a fold of skin with half a syringeful of the spinal cord of the rabbit, which had died of rabies. It had been preserved for 15 days in a flask of dry air. In the following days, fresh inoculations were made. I thus made 13 inoculations. On the last days, I inoculated Joseph Meister with the most virulent virus of rabies.[2]

Joseph Meister was given the 13 injections over a period of ten days. Initially he was given weak preparations whose potency was assessed by simultaneously injecting the preparations into the brains of healthy rabbits. The injections given on the first five days were avirulent, and those given on the subsequent days were increasingly virulent. But Joseph Meister survived. He did not contract rabies, either from the dog bites or from Pasteur's vaccine.

There was an interesting footnote to the story. Many years later, a grateful Joseph Meister became the caretaker of the Pasteur Institute, a duty he performed with exceptional dedication. On 14 May 1945, when Paris was overrun by German forces, Meister, now an old man, was ordered to surrender the keys to Pasteur's tomb by a group of soldiers. Rather than comply with an order which he felt would betray his saviour, Meister refused, went up to his small room and committed suicide.

Greatly encouraged by the success of his vaccine, albeit in only one patient, Louis Pasteur opened a rabies-vaccination clinic next to his laboratory in 1885. He had greatly underestimated the response: as news of the vaccine and a possible cure for rabies spread, large numbers of people from all over Europe, and even some from America, began arriving at his clinic. Most were desperate souls; many had been bitten by rabid animals or had relatives who had been bitten, all demanding the vaccine that would save them from certain death. Louis Pasteur was their only hope. Pasteur and his team of physicians did their best. Records from the period indicate mixed results: some were saved by the vaccine, but it appeared to be ineffective for others. Then one day a young patient entered the clinic. Her name was Louise Pelletier.

Young Louise was brought to Pasteur's clinic by her distraught parents. She had been bitten by a rabid dog 37 days previously. At first Pasteur refused, explaining that she had presented too late. He knew from experience that the disease had taken hold and that after such a long period the vaccine was unlikely to work. But, as with Joseph Meister, Pasteur was put under pressure by the child's parents. Perhaps because he had lost his three beloved daughters and knew what agonies lay before them were they to lose their child, or perhaps because of the memory of his success with Joseph Meister, Pasteur finally gave in and agreed to vaccinate the girl. But this time the story did not have a happy ending. Pasteur's instinct had been right. The girl had arrived too late. Louise Pelletier died of rabies, and those in the medical establishment who had been reluctant to endorse Pasteur's vaccination now accused him of gross negligence. They argued that no one knew how Pasteur's vaccine worked and that his methods were crude and unscientific.

Chapter 8

THE GREAT WHITE PLAGUE

August 1890. Berlin was hosting the World Congress of Medicine, attended by some 7,000 doctors. The opening ceremony was attended by the kaiser and the inaugural speech was given by the eminent German physician Rudolf Virchow. Then it was the turn of Robert Koch.

Koch's speech began with a long, often rambling review of the current state of bacteriology. The audience, expecting more, soon began to get restless, with some even starting to get up and leave. Then Koch began speaking about tuberculosis.

*

Tuberculosis, also known as phthisis, consumption and the 'great white plague', is probably responsible for more deaths than any other infection known to man. It has been called the 'graveyard cough' and the 'robber of youth'. In the eighteenth and nineteenth centuries tuberculosis was endemic in Europe. It is estimated that it was responsible for around one in seven of all deaths.[1] In 400 BC, Hippocrates gave a vivid description of it in his book *Of the Epidemics*:

Early in the beginning of spring, and through the summer, and towards winter, many of those who had been long gradually declining, took to bed with symptoms of phthisis... Many, and, in fact, most of them died, and of those confined to bed, I do not know of a single individual who survived for any considerable time... Consumption was the most considerable of the diseases which then prevailed, and the only one which proved fatal to many persons. Most of them were affected by these diseases in the following manner: fevers accompanied with rigors, constant sweats, extremities very cold, and warmed with difficulty; bowels disordered, with bilious, scanty, unmixed, thin, pungent and frequent dejections. The urine was thin, colourless, unconcocted or thick, with deficient sediment. Sputa small, dense, concocted, but brought up rarely and with difficulty; and in those who encountered the most violent symptoms there was no concoction at all, but they continued throughout spitting crude matters.[2]

Tuberculosis mainly affects the lungs, but the disease can spread throughout the body, affecting many other organs. Common pulmonary symptoms include a cough, which can contain blood, night sweats and fever. Patients also lose weight. Most individuals who suffer from tuberculosis are asymptomatic, around 10 per cent going on to develop an active disease, which if left untreated has a case fatality ratio of around 50 per cent.[3] In 15–20 per cent of those suffering from active disease, the infection becomes extrapulmonary – it goes beyond the lungs – and can spread to other parts of the body.[4] These patients are usually but not always immunocompromised. The infection can spread to the pleura, the brain and spinal cord, the lymphatic system, bones and joints and the genitourinary system. Miliary tuberculosis is a disseminated form of the disease in which the infection progresses unchecked throughout several organs, giving rise to small foci that have the appearance of millet seeds.

Robert Koch had seen many consumptives in his practice. He had watched helpless as the pale, thin, fragile souls, suffering with a sense of foreboding and melancholy, lingered on until death released them from their burden.

Tuberculosis was romanticized by many writers of the time. The poet John Keats, his mother and his brother all succumbed to the disease at an early age. Some thought that consumption conferred upon women a strange attractiveness, one writer describing how it gave these pallid, delicate sufferers a 'terrible beauty'.[5] Edgar Allan Poe's young wife, Virginia, when infected with tuberculosis was described by the writer's biographer as 'delicately, morbidly angelic'.[6] But in truth it was a long, protracted disease, and patients often suffered for years before finally yielding to the *coup de grâce*.

But Koch had made great progress. In 1882 he had announced the isolation of the causative agent of tuberculosis: the first time a human disease was shown to be caused by a microorganism. One year later the bacillus was named: *Mycobacterium tuberculosis*. He had isolated the bacillus from the sputum of patients, and from infected animal tissue, and, by injection, transmitted the disease to guinea pigs. Following these discoveries, Koch formulated, in 1884, his famous postulates: four criteria that had to be satisfied before a microorganism could be deemed to be the agent responsible for a disease. The first postulate stated that the microorganism must be present in all cases of the disease. The second required that the microorganism be isolated from the diseased host and grown in pure culture. The third asserted that the microorganism from the pure culture must, when inoculated into a healthy, susceptible laboratory animal, cause the same disease, and the final postulate required that the microorganism be reisolated from the new host and shown to be the same as the original pathogen.

As we have already seen, Koch preferred to work alone, often spending long hours in the laboratory. He injected guinea pigs with the live tuberculosis bacilli he had finally succeeded in growing in a pure culture. When he gave the animals a second injection he noted after a day or two a reddening of the skin, followed by a vigorous inflammatory reaction at the injection site. He had not seen this reaction after the first injection. This was something unique to the second challenge. Koch realized that this skin response, indicating previous exposure to tuberculosis, could be used as a diagnostic test. If a man or a farm animal had been infected with the tubercle bacillus, the appearance of a skin reaction upon injection would confirm exposure. Koch asked himself what could be happening here. Could this heightened reaction be an indication of immunity? Could he dare to hope that this observation would eventually lead to a treatment?

A few weeks later Koch prepared a glycerol extract of the bacillus, which he subsequently called 'tuberculin'. When he repeated the guinea-pig experiment with tuberculin, he found that the extract also elicited the strange delayed skin reaction. But, more importantly, it also conferred protection in guinea pigs challenged with live bacillus. Repeating the experiment, this time in mice, he again found that it conferred protection. Tuberculin, he surmised, had generated a protective immune response against tuberculosis. The extract also appeared to be safe, at least in guinea pigs and mice.

At the International Medical Congress in Berlin in the summer of 1890, Robert Koch, the father of bacteriology, finally began to speak about his latest work on tuberculosis. To the delight of his audience, he announced that he had finally found a substance that could arrest the growth of the tubercle bacilli both *in vitro* and *in vivo*. Injection of the substance – he didn't go into details about its nature – into guinea pigs, he claimed, rendered healthy animals resistant to tuberculosis, and when injected into animals already suffering from tuberculosis it arrested the disease and prevented it from spreading throughout their bodies.

The news was electrifying. Shortly after the congress the German medical establishment approached Koch. They suggested immediate clinical testing. The kaiser was interested, they told him. He must proceed with haste, for the glory of the fatherland. But Koch was hesitant. It was too early to be thinking about clinical tests, he said. Tuberculin had never been tested on a human subject. In the end, Robert Koch decided that he would himself become a guinea pig by injecting the extract into his own body. But he kept the self-experiment secret. Three or four days after giving himself an injection of tuberculin, Koch became very unwell. His breathing became laboured and his joints became inflamed. He began to vomit and was seized by rigors. His temperature rose to 40°C. His wife begged him to call a doctor. Koch refused. He wanted to study the effects of tuberculin and was keeping detailed records of all his symptoms. After several agonizing days his symptoms abated, and Koch made a full recovery. He decided that it was now safe to begin clinical trials.

The trials began at the Charité hospital in Berlin. Koch restricted his vaccination to milder cases, but the news soon began to spread. On hearing that there might be a vaccine against the dreaded disease, thousands of consumptives flocked to Berlin. Initial results were promising, and Koch was hailed as a saviour and a hero. Many felt that his new vaccine would eliminate the dreaded disease from Europe. Koch was given the Freedom of Berlin and the Grand Cross of the Red Eagle from the kaiser. As he sat at home one evening, a telegram arrived. Koch noticed that it had come from Paris. His wife and daughter gathered round as he read it. 'Congratulations, Louis Pasteur.'

But what exactly was tuberculin? At the Berlin Congress Koch had referred to it as 'lymph', not revealing its true nature or composition. Later, tuberculin was revealed as a filtrate of the tubercle bacillus grown for six to eight weeks in glycerol medium and evaporated into one tenth of its volume.

Sadly, tuberculin therapy did not live up to its promise. It wasn't to be a universal panacea for the white plague. Although some people clearly benefited, gaining protective immunity, many didn't. Some patients given tuberculin succumbed to tuberculosis and died. In January 1891, the bodies of 21 patients who had died after receiving Koch's tuberculin were autopsied by Rudolf Virchow, who found that the disease had invaded several internal organs. Virchow observed the 'millet seeds' everywhere. Miliary tuberculosis. 'These cases of phthisis are worse than any I have seen in my lifetime,' said Virchow. Distraught, Koch finally acknowledged that he had been too hasty in offering the vaccine to patients. It later emerged that even the animal work hadn't been carried out thoroughly. Under relentless questioning, Koch confessed that he had not autopsied any of the guinea pigs used in his original experiments.

But tuberculin still worked for some, and the tuberculin test was widely used diagnostically. In 1905 Robert Koch was awarded the Nobel Prize in Physiology or Medicine for his investigations and discoveries in relation to tuberculosis.

How did these vaccines work? What were the underlying mechanisms? At that time no one had any idea of the existence of an immune system or which elements in the body conferred protection, and it would take a Russian biologist to unravel some of the mysteries of our defence system.[i]

---

i   Tuberculosis presents the immune system with a seemingly insurmountable problem. The tubercle is a canny enemy which has evolved a means of surviving inside the very cell that is supposed to destroy it: the phagocytic macrophage. The tubercle bacillus lives inside, resisting phagocytosis, and, like an unwanted lodger, fights all attempts by the host to remove it, remaining within the cell and then erupting when the host is weak to wreak havoc within the body. The immune system tries many strategies: as it is an intracellular pathogen, antibodies are pretty much useless, and the only defences are the cell-mediated immune responses, but these are also often ineffective. The immune system must finally resort to the strategy of all invaders who come across a stubborn enemy that has ensconced itself inside a fortified city: it lays siege.

Chapter 9

THE ROSE THORN AND THE STARFISH

Messina, Sicily, 1882. Dawn was breaking on a windswept, deserted seashore. The waves on the strait curled up like giant blue curtains, unfolding then rolling into quiescence. A solitary figure was pacing up and down the shore; dressed in black, the man, his long dark hair flowing wildly in the wind, appeared to be drunk. But he was not. The man was oblivious to his surroundings. His mind was turbulent, his thoughts racing like a thousand wild horses galloping down the plains of his birthplace, the Russian Steppe.

Élie, or Ilya, Metchnikoff was born in the village of Ivanovna, Kharkov, in 1845. A brilliant student who completed his four-year degree in Natural Sciences in two, Metchnikoff trained as a zoologist, and studied the digestive processes of invertebrates. But Metchnikoff was unstable, had a rather fiery temper and was given to frequent bouts of depression. Before his landmark discoveries in immunity he had suffered several breakdowns and had even attempted suicide. His last breakdown had been precipitated by being dismissed from his post at the University of Odessa for political activism.

In 1882, one year after Pasteur's trial at Pouilly-le-Fort, Metchnikoff, along with his young wife and three daughters, set up a marine-biology laboratory in Messina. There he continued his research into invertebrates, which included microscopic examinations of starfish larvae. Noticing some mesenchymal cells within the bodies of the transparent larvae, he wondered about their function. At first, he thought that the cells were involved in the digestive process, then considered whether they had some protective function. These deliberations led to his most famous experiment.

He wrote:

One day when the whole family had gone to a circus to see some extraordinary performing apes, I remained alone with my microscope, observing the life in the mobile cells of a transparent star-fish larva, when a new thought suddenly flashed across my brain. It struck me that similar cells might serve in the defence of the organism against intruders. Feeling that there was in this something of surpassing interest, I felt so excited that I began striding up and down the room and even went to the seashore to collect my thoughts.

I said to myself that, if my supposition was true, a splinter introduced into the body of a star-fish larva, devoid of blood-vessels or of a nervous system, should soon be surrounded by mobile cells as is to be observed in a man who runs a splinter into his finger. This was no sooner said than done.

There was a small garden to our dwelling, in which we had a few days previously organized a 'Christmas tree' for the children on a little tangerine tree; I fetched from it a few rose thorns and introduced them at once under the skin of some beautiful star-fish larvae as transparent as water.

I was too excited to sleep that night in the expectation of the result of my experiment, and very early the next morning I ascertained that it had fully succeeded.

That experiment formed the basis of the phagocyte theory, to the development of which I devoted the next twenty-five years of my life.[1]

After setting up his experiment, Metchnikoff retired to bed knowing that the reaction would take at least 12 hours. As dawn broke he looked down the microscope and found that the mobile mesen-chymal cells of the starfish larva had surrounded the rose thorn. These were defensive cells, and they were attempting somehow to surround and destroy the thorn. This was the beginning of his phagocyte theory.

News of his finding rapidly spread beyond Messina and reached the ears of the doyen of German medicine, Rudolf Virchow, who happened to be visiting Sicily at the time. On a warm day that summer, Virchow turned up unannounced at Metchnikoff's house. Metchnikoff excitedly showed his visitor the slide. Virchow was impressed, but he questioned Metchnikoff's interpretation, suggest-ing that the parasite was merely invading the cells. Inflammation, he said, is always detrimental. However, he encouraged the excitable young man to publish his findings.

Metchnikoff named his cells 'phagocytes', in Greek 'the devour-ing ones', believing that their role was to ingest and then digest foreign invaders. He considered them to be the first line of defence against pathogens, and carried out experiments on rabbits show-ing that the phagocytic macrophages in animals vaccinated and challenged with anthrax bacilli were more active and congregated around the bacteria in large numbers. In comparison, unvaccinated rabbit macrophages appeared unenthusiastic, sluggish even. His findings came to the attention of Louis Pasteur, who felt sure that Metchnikoff had unwittingly stumbled upon the mechanism underlying vaccination. In 1888, Metchnikoff was offered and took up a post at the Pasteur Institute, where he remained for the rest of his career.

The discovery of phagocytosis completely changed Metchnikoff. He became a driven man with only one solitary goal, one purpose, one destiny that he had to follow to the very end: to understand the function of his beloved phagocyte. It was as if a beam of light had dispelled the darkness, illuminating in an instant the mystery of how the body fought off invading pathogens. Until then, no one had had any idea how the body combated contagion. Metchnikoff thought he knew how: it was the phagocyte that undertook this task. He plunged into his research with renewed energy, forgetting his melancholy, the bouts of depression that had nearly driven him to suicide.

A substantial body of work followed. Metchnikoff described how phagocytes captured living organisms, showing this to be an active process. He showed that the ingested pathogen was contained within a 'vacuole', a bag containing toxic chemicals within which the trapped parasite became digested by acidification and enzymes derived from the leucocyte.

He studied the inflammatory process showing it to be, contrary to thinking at the time, a beneficial response. He described how certain types of white blood cells found in the blood stream squeezed themselves between the single cell layer lining tiny blood vessels, the capillaries, to enter tissues to carry out phagocytosis. He had, in fact, described neutrophils, which he called 'microphages', distinguishing them from macrophages. 'On macrophages and microphages [neutrophils],' Metchnikoff writes, 'I suggest calling all elements macrophages which generally possess a simple nonpolymorphic nucleus which is round or frequently oval... as microphages I call smaller amoeboid cells which can be easily stained, with a largely polynuclear and fragmented nucleus and faint protoplasm.'[2]

Metchnikoff demonstrated that there was enhanced phagocytosis when the macrophages were exposed to bacterial products. This activation of macrophages has now been shown to be due to various microbial substances binding to receptors – now called 'toll-like receptors' (TLR) – found on the surface of the phagocytes. Metchnikoff showed that macrophages were also scavengers, taking up and digesting dead and senescent cells.

At the Pasteur Institute, Metchnikoff and his colleagues studied the course of the tuberculosis infection. They described the formation of granulomas: strange, large masses made up of giant cells, formed by fusion of macrophages and other cells. The granulomas were the prison-castles of the immune system, walling off pathogens and preventing their dissemination throughout the body.[i]

On the role of the macrophage in tuberculosis, Metchnikoff wrote:

> The real phagocytes (in particular macrophages) serve as critical defenders of the host against tubercle bacilli... without doubt the phagocytes are capable of engulfing live and virulent tubercle bacilli actively... and the giant cells are capable of killing these parasites... of course, it is possible that tubercle bacilli can evade the deadly activity of giant cells.[3]

Metchnikoff's work led to the belief that the mechanism of immunity had been discovered. It was an enhanced phagocytosis that led to immunity following vaccination.

But in 1901 Metchnikoff read reports of an opposing theory of immunity, one that didn't appear to involve cells at all. Strange substances in the blood were, it seemed, also capable of conferring immunity. This was the beginning of the 'humoral theory' of immunity. Substances in the humours, including blood, rather than cells were the effectors of immunity. Thus, the stage was set for one of the most enduring battles of immunity: the debate between the humoral school and the cellular school, one that would rage unresolved for another three decades.

Phagocytosis, as discovered and described by Metchnikoff, is an example of an innate immune response. This natural immune system, as we shall see later, lacks the specificity of the adaptive immune response, which by generating the heightened secondary response confers protection following vaccination.

When a pathogen invades, the body attempts to get rid of it by using innate, non-specific defences and, failing that, by specific adaptive immune responses. In practice, as we shall see in Chapter 27, the two systems are interrelated and work together to quell the invasion.

*

So what happens when a foreign body breaches our outer defensive wall, the skin, and manages to enter deeper tissue? Let's look at how the body responds to such an insult.

Imagine that when Metchnikoff was picking rose thorns to inject into the starfish larva, in his excitement he pricked himself with a thorn. The thorn is now sticking out of his finger. To defend our hero from the microbes coating the thorn, the body's inflammatory response kicks in. This inflammatory response is a major part of the innate immune response. But what exactly is the inflammatory response? What are the key players, how does it begin and what are the consequences?[ii]

The word 'inflammation' is derived from the Latin *inflammare*, to set on fire. The Roman writer Aulus Cornelius Celsius first described the four cardinal signs of the inflammatory response in the first century AD: heat, redness, swelling and pain.[4] Sure enough, Metchnikoff's finger would have been red, hot and swollen. The first three signs are due to mediators, chemicals produced by the body (this was not known during Metchnikoff's time), causing the dilatation of blood vessels, which also become more leaky.[iii] This vasodilatation causes more blood flow, giving rise to warmth and redness. The increased leakiness, capillary permeability, results in fluid leaking out of the vessels, giving rise to swelling, oedema. The fourth is due to mediators stimulating pain receptors. Galen, physician to the Roman emperor Marcus Aurelius in the second century AD, added a fifth sign: loss of function (*functio laesa*). Metchnikoff couldn't bend the swollen and painful finger; couldn't use it to hold his pen: loss of function!

Galen also proposed that during inflammation blood filtered from vessels into the tissues. But it took the invention of the microscope in the sixteenth century to show this process. Henri Dutrochet, the French physician–physiologist, described in 1824 how white blood cells stick to the walls of the blood vessels before migrating into the inflamed tissue.[5] This is called 'margination'. In the nineteenth century, Wagner and Cronheim described how these white cells rolled on the vessel walls, then squeezed between the endothelial cells and entered the inflamed tissue.[6]

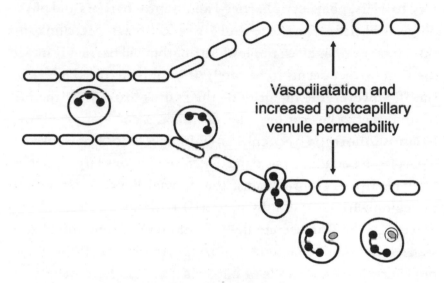

Vasodilatation and increased postcapillary venule permeability

Figure 4. Inflammation and phagocytosis.

Why did these white blood cells leave the blood and enter the tissue? It was, as we saw earlier, Ilya Metchnikoff who provided the answer: phagocytosis (Fig. 4). They were getting into the tissues to engulf and destroy the intruders.[iv] When these phagocytes set about their task they release 'lytic' substances, which can also damage surrounding healthy tissue. The accumulation of dead cells, digested material and fluid forms a substance called pus.

Metchnikoff's finger would have become swollen. If he incised the swelling the pus would come out and he would have found neutrophils in the yellow fluid. These cells are characteristic of an acute inflammatory response. Acute inflammation can last from minutes to days, but there is also another type of inflammation which lasts a lot longer. This is chronic inflammation, seen for example in a joint inflamed by rheumatic arthritis; it can last from weeks to years and has an oedematous fluid mainly composed of lymphocytes and macrophages.

The innate defences can be thought of as barriers: inflammatory barriers, phagocytic barriers, anatomical barriers and physiological barriers. The rose thorn would breach Metchnikoff's skin. Intact skin is an example of an anatomical barrier. If intact, the skin surface seems to be rarely, if ever, penetrated by bacteria. If, however, the integrity of the skin is broken (by the bite of an insect, a laceration, a needle stick, a cut – or penetrated by a rose thorn), pathogens can gain access to deeper tissues. Apart from being a physical barrier, skin contains large numbers of 'friendly' microorganisms, the normal flora, which inhibit the colonization of skin by potential pathogens. Sweat glands present in the skin secrete fluid containing antimicrobial substances that inhibit bacteria. Lysozyme, one such substance, can break down the cell walls of bacteria. We also have 'sebaceous glands' on our skin. These are found in the hair follicles and produce an oily secretion called 'sebum', which contains lactic acid and fatty acids, which maintain the pH of the skin at a level between three and five, which is inhibitory to the growth of most pathogens. However, some bacteria can metabolize sebum and live as commensals on the skin, sometimes even causing a severe form of acne.

Another important anatomical barrier is provided by the mucous membranes which line the conjunctivae of our eyes, as well as the alimentary, respiratory and urogenital tracts. Although many pathogens can enter the body by penetrating mucous membranes, several non-specific defence mechanisms attempt to prevent this mode of entry; for example, saliva, tears and mucous secretions can wash away potential invaders. They also contain antibacterial or antiviral substances such as lysozyme. A viscous fluid called mucus, secreted by the epithelial cells of mucous membranes, can entrap foreign microorganisms. In the respiratory tract, the mucous membrane is covered by cilia, hair-like protrusions which are situated on epithelial-cell membranes. These cilia move synchronously, propelling mucus-entrapped microorganisms away from the tracts.

In addition, as on the skin, normal flora can colonize the epithelial cells of mucosal surfaces. These harmless microorganisms tend to outcompete pathogens for attachment sites on the epithelial-cell surface and for important nutrients.

Examples of physiological barriers include body temperature, low pH and a whole slew of soluble factors such as cytokines, complement and acute-phase proteins.[v] We will come across these factors again in Chapters 12, 13 and 23.

Many species are resistant to certain diseases simply because their normal body temperature inhibits the growth of the pathogens. For example, chickens have innate immunity to cholera because their high body temperature inhibits the growth of the cholera vibrio. Gastric acidity is another important physiological barrier: few ingested microorganisms can survive the low pH of the stomach. One reason that newborns are susceptible to some diseases that do not seem to cause problems in adults is that the pH of their stomachs is less acidic than those of adults.

Once the inflammatory response has subsided, phagocytes will clear away the tissue debris. This is followed by tissue repair and regeneration of new tissue. There can be several outcomes of an episode of acute inflammation. The best outcome is resolution, but often there is scarring. Unfortunately, sometimes a walled-off collection of pus forms an abscess. Rarely, unresolved acute inflammation can progress to chronic inflammation.

i In tuberculosis the immune system tries to wall off the infection by surrounding the infected macrophage, forming a collar of cells around it, consisting of fused macrophages, giant cells, T lymphocytes, B lymphocytes
and fibroblasts. This structure is called a 'granuloma'; hence tuberculosis
is called a 'granulomatous disease'. In the centre of the granuloma there
can be cell necrosis, giving the tissue there a cheesy-looking appearance:
this is thus called *caseous necrosis*, the hallmark of the tuberculous lesion.
The granuloma not only prevents the spread of the disease but also allows
for cell-to-cell communication, vital in generating an effective immune
response against the pathogen. Inside these granulomas the immune
system gets to work.

ii The inflammatory response begins when a component of a microbe,
such as lipopolysaccharide (LPS) found in the cell walls of some bacteria,
interacts with cell surface receptors (toll-like receptors, TLR) found on the
surfaces of sentinel cells of the immune system. These sentinel cells include
macrophages and dendritic cells. Binding of microbial ligands by TLRs
sends signals to the interior of the cell, which, by causing the expression
of genes, leads to the production of key proteins such as cytokines that
modulate immune responses.

iii The twentieth century saw the discovery of the substances that drive the
process of inflammation, the chemicals that cause the blood vessels to
become more permeable, more dilated, that draw the phagocytes away
from the blood onto the tissue that constricts smooth muscles: the *mediators* of inflammation.

Mediators can be cell-derived or plasma-derived; the latter are mainly
made in the liver. Cell-derived mediators include arachidonic acid deriva-
tives, cytokines, vasoactive amines and the various antimicrobial substances
found within lysosomal vesicles. Plasma-derived factors include the proteins of the clotting, complement and kinin systems.

Let's look at some of these mediators in a bit more depth. One of the
key vasoactive amines and a key mediator of the inflammatory response
is histamine, a chemical released by cells in response to tissue injury. The
binding of histamine leads to dilatation and increased permeability of

blood vessels. Kinins, small peptides normally present in plasma in an inactive state, also cause dilatation and increased permeability of capillaries following activation as a result of inflammation and injury. One such kinin, bradykinin, also stimulates pain receptors in the skin. The injured tissue may contain damaged blood vessels, and enzymes of the clotting cascade that arrive at the area will attempt to stem the blood loss by deposition of a clot composed of insoluble fibrin. The fibrin strands also have the effect of walling off the injured area from the rest of the body, thereby preventing the spread of infection.

iv The final barrier is the phagocytic barrier. We now know that macrophages, neutrophils (also described by Metchnikoff) and eosinophils are phagocytes. These cells ingest bacteria by causing the cytoplasm to flow around the pathogen, resulting in its engulfment. The engulfed pathogen is then taken inside the cell, where it is enclosed in a membrane-bound vacuole called a phagosome. Bags of chemicals called lysosomes containing various antimicrobial substances, such as hydrogen peroxide, then bind and merge with the phagosome to give rise to a structure called the phagolysosome. Inside the phagolysosome, digestion of the pathogen occurs. Reactive oxygen species within the phagolysosome cause destruction of the pathogen by a series of chemical reactions collectively called the oxidative burst. Macrophages show a greater oxidative burst compared to neutrophils.

v There are several important cytokines that are part of our physiological barrier. For example, interferon gamma can inhibit viral replication. Interleukins 1 and 6 can raise body temperature, which again can inhibit pathogens. Interleukin 8 can attract neutrophils to the site of the infection, and the acute-phase protein CRP can activate the complement system, which, by a series of reactions, can lyse bacterial cells, generate inflammation and coat bacteria, facilitating their phagocytosis.

Chapter 10

THE STRANGLING ANGEL

Émile Roux was about to leave the hospital for the day when the small child was brought into the ward. Something about this child, a boy who looked around four years old, made Roux – a physician who was used to seeing death and disease daily – stay and observe. The child was complaining of a painful throat. He looked feverish. 'Respiratory diphtheria,' Roux muttered to himself. 'Probably caught it a few days ago.'

Over the next few weeks the boy developed a swelling in the neck, and the arrival of this dreaded 'bull neck' heralded a grave prognosis. Roux examined the child and noted the grey/yellow coating inside his throat, the pseudomembrane characteristic of the infection. The boy, his complexion a pale blue, began to cough. Roux knew that soon he would have great difficulty swallowing and that his fever would continue unabated. Not long after, the child developed a bloody nasal discharge. Roux thought it likely that he would develop complications. His heart would become inflamed; his diaphragm might suddenly be paralysed.

There was nothing he could do for the boy. The child would soon die, probably of respiratory failure. He knew that diphtheria in children killed nearly half of those infected. Of those who survived, many were left paralysed and with greatly weakened hearts. The yellow, dirty-looking membrane in the throat often choked the victims, causing their deaths.

*

After Pasteur's second stroke in 1888, Émile Roux took over as head of the Pasteur Institute. Now he was finally able to step out of his great mentor's shadow and pursue his own interests. He decided to wage war on diphtheria: 'the strangling angel'.

Roux and his colleague, the physician–bacteriologist Alexandre Yersin, had found club-shaped bacteria in the pseudomembranes, bacteria that had been shown in 1884 by Theodor Klebs and Friedrich Loeffler to be the causative agent of diphtheria.[1] But Roux wondered how these bacteria, residing in the throat, caused the disease. Did the bacteria spread everywhere, or were they producing some poison that led to the effects seen throughout the body?

He isolated bacteria from the throat of a sick child, grew them for four days in broth culture and passed the broth through a porcelain filter. Roux then injected the amber-coloured clear filtrate, free of microorganisms, into guinea pigs. The animals did not get diphtheria. This disappointed Roux; he had expected to find the poison in the filtrate. He grew the bacteria in liquid culture for a longer period, 42 days, and repeated the experiment. When he injected this filtrate into guinea pigs they all died. Roux had demonstrated that diphtheria was caused by a poison. There was no need for the bacillus to be present to cause the disease.[2]

*

Meanwhile, in Berlin, two medical scientists were also studying diphtheria: the German physiologist Emil von Behring and Kitasato Shibasaburo, a Japanese physician from Tokyo Imperial University.

Behring was born in 1854 at Hansdorf, in Prussia. As the son of a schoolmaster, one of 13 children, he did not have the means to secure a place at one of Germany's traditional universities, and in 1874 he entered the Army Medical College in Berlin. He obtained his medical degree in 1878 and, as required, spent a few years working for the German army. In 1888, he obtained a position as assistant to Robert Koch at Berlin's Imperial Hygienic Institute.

In 1890, Behring and Kitasato began working on diphtheria, conducting a series of experiments that involved the inoculation of rats,
guinea pigs and rabbits with a weakened form of the diphtheria bacillus. As expected, the animals remained resistant to the disease when
exposed to the live bacillus. Then they took some immune serum – the
straw-coloured liquid separated from blood after it had been allowed
to clot – from these animals and injected them into a group of unimmunized animals who had previously been infected with the fully
virulent bacilli. To their surprise, these animals, who were expected
to succumb to the infection, also remained resistant to diphtheria; the
injected immune serum had protected them from the disease (Fig. 5).[3]

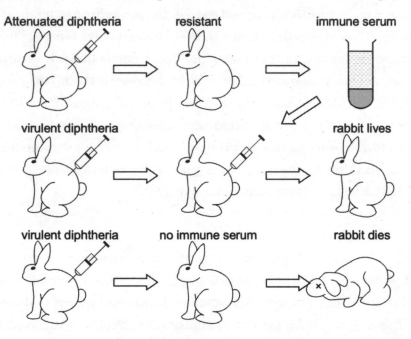

Figure 5. Serum immunity against diphtheria.

In a subsequent experiment, Behring and Kitasato showed that
animals who would normally yield to the diphtheria toxin, recently
discovered by Roux and Yersin, would survive if the toxin preparations were mixed with immune serum, left for 24 hours and then
injected into the animal. Mixing the toxin with the immune serum,

it appeared, was neutralizing the toxin. Also, if the immune serum was incubated with living bacteria and injected into healthy animals, they did not contract the disease.[4] Something in the serum was protecting these animals from the toxin, and this serum factor could be transferred from one animal to another, conferring protection when challenged by the toxin or by the bacteria. Behring called this serum factor 'antitoxin'. Today we call it an 'antitoxin antibody'.

Behring began producing antitoxin. He did this by injecting sublethal doses of live or killed cultures of diphtheria bacilli into animals, removing blood and then separating out the serum fraction containing the neutralizing antibody. By this method Behring was able to produce antitoxin capable of protecting animals from lethal doses of bacillus or toxin. Later he was able, using a similar method, to obtain protective antitoxin against tetanus toxin.

Behring felt he was on the verge of a breakthrough against the strangling angel. He approached Robert Koch, who, to his surprise and annoyance, advised against publication of his findings. 'You need more data,' he said. Behring was incensed. 'If we don't publish, Roux will. The French have been working on the same problem for a long time,' he argued.

Then, serendipitously, Behring encountered a patient, a little girl who was dying of diphtheria. He decided to act. On Christmas Eve 1891, he treated the child with antitoxin prepared from an immunized sheep. He had not obtained permission from Koch. In just a few hours, almost miraculously, the girl began to recover. During the New Year's Eve celebration held at the institute, Emil von Behring announced the outcome of his human experiment: the girl had been saved by a novel treatment. He called it 'serum therapy'.

Serum therapy, or 'serotherapy', was enthusiastically adopted by a desperate medical fraternity, and the method spread across Europe and eventually to the rest of the developed world.

On 15 October 1894, a boy aged eight who lived in Lewes, Sussex, became the first patient to be treated for diphtheria with sera manufactured in England. The antitoxin was obtained by the bleeding of a horse which was being immunized by Sir Charles Sherrington, a leading neurophysiologist of the time. Here is Sherrington's own account of what happened.

Ruffer is almost as French as English. Burchard the eminent Paris physician was a brother-in-law. Ruffer was often in Paris and constantly brought us news of the Institute Pasteur. Returning from one of these visits he spoke of the treatment of diphtheria which was being tried out there. Injection of the serum of unimmunized horse. He would like us to try it. I had a spare stall in the stable at my veterinary hospital (at the Brown Institution south of the Thames) and we got a horse and began inoculating it with gradually increasing doses of diphtheria cultures. We had been at this [a] week or two, and had serum at least partially effective in guinea pigs. We were badly in the dark about dosage.

Then oddly enough, one Saturday about seven in the evening, came a telegram from my brother-in-law in Sussex. 'George has diphtheria.' George, a boy of eight, was the only child. The house, an old Georgian house, was some three miles out of Lewes, set back in a combe under a chalk down. There was no train that night. I did not at first give thought to the horse and when I did regretfully supposed it could not yet be ripe for use. However, I took a cab to find Ruffer – no telephone or taxi in those days. Ruffer was dining out, I pursued him and got a word with him. He said, 'By all means you can use the horse but it is not yet ripe for trial.' Then by lantern light at the Brown Institution I bled the horse into a great 4L flask duly sterilized and then plugged. I left it in ice to settle. Then after sterilizing smaller flasks and pipettes drove home to return at midnight and decant the serum and sterilize needle syringes. By Sunday's early train I reached

Lewes. Dr Fawsset of Lewes was waiting at the railway station in a dog cart. I joined him, carrying my awkward package of flasks and such like. He said nothing as I packed them into the dog cart but when I climbed up beside him he looked at me. 'You can do what you like with the boy; he will not be alive at tea time'.

We drove out to the old house. Tragedy seems to shroud it. The boy was breathing with difficulty. He did not know me. The doctor helped with injecting the serum. The syringes were small, and we emptied them time and again. The doctor left. Early in the afternoon the boy was clearly better. At 3 o'clock I sent a messenger to the doctor. Thenceforward progress was uninterrupted. On the Tuesday I returned to London and sought out Ruffer. His reaction was that we must tell Lister. The great surgeon had visitors – some continental surgeons – to dinner. 'You must tell my guests,' he said, and he insisted.[5]

In an earlier account of this dramatic, episode Sherrington added that 'the boy had severe paralysis for a time. [But] he grew to be six feet and had a commission in the First World War.'[6]

Behring, Kitasato and others had discovered antibodies, strange molecules in the blood of animals that had the capacity to neutralize bacterial toxins. Serotherapy became a widely used method in the treatment of diphtheria, and in three years around 20,000 children were vaccinated in Germany using sheep and goat antiserum.[7] In 1891, Behring began immunizing horses, and horse antiserum was used in mass immunizations.

But, as with Koch's tuberculin, serotherapy wasn't a universal panacea. Clearly, a large number of patients were saved from diphtheria, but there were also deaths, and many suffered from strange reactions. Often trials had to be stopped.

In the early 1900s, prevention and treatment of diphtheria – and subsequently of tetanus – involved giving patients horse antiserum containing mixtures of toxin and antitoxin. But Alexander Glenny, a researcher at the Wellcome Laboratories in London, made a chance discovery which two decades later led to the use of a safer prophylactic alternative in humans. He found that a batch of diphtheria toxin had become less toxic yet retained its immunogenicity – its capacity to induce an immune response. Investigating why this batch had lost its potency, Glenny discovered that the toxin had been stored in large butts that had been chemically sterilized with formalin. Further experiments by Glenny found that the formalin had converted the toxin into a harmless toxoid. Glenny in London and Ernst Löwenstein in Vienna began using toxoids, made by treating toxin with formalin, to immunize horses.[8]

The other infection that was treated with serotherapy was tetanus. Kitasato Shibasaburo, working in Robert Koch's laboratory, had isolated and grown the tetanus bacillus, *Clostridium tetani*, in pure culture, and together with Behring had produced an antitoxin effective against the tetanus toxin. Like diphtheria, the pathology of tetanus involved its deadly toxin, tetanospasmin, produced by the bacillus. Kitasato was able to reproduce tetanus in animals by injecting a pure culture of bacillus or a toxin preparation.[9]

The incubation period of tetanus – the time it takes from infection to the appearance of signs and symptoms – is around a week, but can be as long as several months. Generally, if the incubation time is short then the symptoms tend to be worse. The most common type of tetanus is generalized tetanus. The first sign is lockjaw, or trismus, where the muscles used for chewing become clamped. The next sign is the unintentional grin, *risus sardonicus*, caused by the uncontrolled contraction of the facial muscles. Then difficulty of swallowing and stiffness of the neck muscles occur. The large muscle groups of the body can go into periodic spasm, causing a painful arched body posture known as opisthotonos, which can sometimes break the spine. This has been illustrated in a famous painting of 1809 by the surgeon Sir Charles Bell, *Tetanus Following Gunshot Wounds*. As the name suggests, it depicts the death agony of a soldier who had contracted tetanus through a gunshot wound in the Napoleonic Wars. The painting shows the contraction of the extensor muscles of the victim's back, causing it to arch like a bow; the patient may get locked into this position. Sufferers may also sweat profusely. Spasms can last for around a month. The respiratory muscles can also spasm, resulting in obstruction of airflow, which can lead to death by asphyxiation. Throughout all this the sensorium – the sensory apparatus of the body – remains intact, adding greatly to the suffering.

Behring and Kitasato's serotherapy helped save the lives of thousands of injured soldiers in the First World War. Behring's fame grew. He received numerous awards and honours, which included the Iron Cross, rarely given to a non-combatant. Then, in 1901, he was awarded the first Nobel Prize in Physiology or Medicine for his work on serotherapy.

Behring spent his later years studying tuberculosis, ironically contracting the disease himself when he was 50. He fought the white plague with all his might, but this was one battle the former military doctor could not win. Behring died from tuberculosis aged 63.

Niels Jerne, whom we shall meet later, paid homage to Behring and Kitasato in his 1984 Nobel Lecture:

Let me first recall some of the essential elements of the immune system, with which I shall be concerned. In 1890, Behring and Kitasato were the first to discover antibody molecules in the blood serum of immunized animals, and to demonstrate that these antibodies could neutralize diphtheria toxin and tetanus toxin. They also demonstrated the specificity of antibodies: tetanus antitoxin cannot neutralize diphtheria toxin and vice versa.[iiio]

i Behring also used a mixture of horse antitoxin and toxin in his early serotherapy, a form of active immunity. Obviously, administering the toxin by itself would have had disastrous results, so another approach is to use toxoids. In this approach, a toxin is modified chemically or by heat, so it is no longer able to bind to the toxin receptor and poison the cell. But it is still immunogenic. So you get a good, effective immune response, in this case an antitoxin antibody response, but no toxic effects. Examples are diphtheria and tetanus toxoids that are still in use today.

ii The antibody response can demonstrate the concept of specificity, a key feature of the adaptive immune response. In general, each antibody binds to a unique antigen that triggered its formation, and no other, so antibodies against tetanus toxin will not bind to diphtheria toxin.

Chapter 11

THE CHEMIST

Strasbourg, 1873. A young medical student sat alone in the small pathology laboratory at the school of medicine. Outside, darkness was falling, and a biting wind whistled through the narrow streets. It was the beginning of winter, and it would be a cold and merciless one. Most of the other students and staff had long departed the cold building for the comfort of their warm fires and dinners, but not this student. There was one other person left in the laboratory, but he too was preparing to leave. This was Professor Wilhelm von Waldeyer, the renowned anatomist, who had discovered the organization of the nervous system, and who had first coined the term 'chromosome' and described the ring of tonsils in the throat: the 'ring of Waldeyer'. Passing the small laboratory, Waldeyer saw the young student though a half-open door, hunched over a microscope. Oblivious to everything around him, the young man was lost in whatever he was observing.

Waldeyer knew the young man well. 'The chemist', they called him. He was the best student in his year, but he was a strange one. He didn't seem to care much for clinical medicine, choosing instead to spend all his free time in the laboratory, staining various tissues and cells with the newly discovered aniline dyes. He was fascinated, almost obsessed, by histology and chemistry, having already published a paper on staining. Waldeyer shook his head and walked away. It would be hours before the porters would finally force the young man to put his beloved slides away and leave the building. Little did the professor know that the young man would go on to make seminal discoveries in immunology, haematology, pharmacology, oncology and bacteriology; that he would discover the first chemotherapeutic agent against an infection, Salvarsan; and that one day he would also receive the ultimate accolade: the Nobel Prize in Physiology or Medicine. His name was Paul Ehrlich.

*

Paul Ehrlich was born into a large Jewish family in Strehlen, Upper Silesia, now part of Poland. His father was a distiller and innkeeper. After an unremarkable secondary education, Ehrlich attended the universities of Breslau, Strasbourg and Leipzig. During his university studies he was greatly influenced by his mentor, Waldeyer.

In 1878 Ehrlich was awarded his doctorate in medicine after submitting a dissertation on the theoretical and practical aspects of tissue staining. Shortly after graduation Ehrlich went to work in Friedrich von Frerichs's clinic at the Charité hospital in Berlin, where in addition to his clinical work he continued his histological research. Ehrlich demonstrated that different types of dye – acidic, basic or neutral – stain distinct parts of cells. Using staining, he was the first to identify several different types of white blood cell: mast cells, lymphocytes and eosinophils. This seminal work would later lead to the development of several staining methods that characterized the field of haematology. In 1882, Ehrlich successfully stained the tuberculosis bacillus that had been discovered by Koch. His method, later modified by Franz Ziehl and Friedrich Neelsen, is still used today. Indeed, it was from this method that the most widely used staining technique in bacteriology, the Gram stain, was subsequently discovered.

Ehrlich was appointed house physician at Frerichs's clinic at the Charité hospital in Berlin, and in 1883 he married Hedwig Pinkus, the daughter of a textile merchant, ten years his junior. In 1888, Ehrlich incidentally discovered tubercle bacillus in his sputum, most likely contracted by the handling of laboratory specimens. Alarmed, he hastily left Berlin with his wife and headed to Egypt, where he recuperated for a year. On his return to Berlin, in 1889, Ehrlich decided to be treated with Koch's tuberculin. He never contracted the disease again. In 1889 Ehrlich set up a small laboratory in a rented flat and began working, unpaid, on plant toxins.

His first forays into immunological research involved using ricin and abrin toxins as antigens. He fed these toxins to young mice at gradually increasing doses. Once he had achieved protection from 'challenge' – that is, once the mice had become immune to the effects of a lethal dose of toxin – he examined the blood and found elevated levels of specific antibodies against the toxins. The mice had developed an active form of immunity.

Ehrlich went on to show that the progeny of female mice thus immunized also showed specific but transient immunity to these toxins. This transient immunity could be sustained by suckling on maternal milk. Ehrlich realized that in this case a passive, albeit specific immunity had been established in these young mice. They had, it appeared, received the protective antibodies via their mother's milk.

In another experiment, Ehrlich took some newborn mice that had been born to an unimmunized mother and allowed them to suckle an immunized mouse. The baby mice received passive immunity: the suckling had enabled the young mice to obtain antibodies from the immunized surrogate mother. In a separate experiment, he found that a normal lactating mouse, if injected with antiserum from an immunized mouse, passed the antibodies passively to its offspring.

On the transfer of protective antibodies to toddlers, Ehrlich wrote:

From our experiments we can conclude that the newborn are endowed with antibodies from the mother... my experiments demonstrate that indeed the milk is capable of transferring antibodies to the suckling toddler and thus provide the toddler with a high immunity which further increases through breastfeeding.[11]

Ehrlich's elegant experiments highlight an important concept of specific immunity: it can be active or passive. In active immunity, the immunity is acquired by the active engagement of elements of the host animal's immune system, leading, for example, to the production of antibodies. So the animal gets an infection and the specific immune system kicks in, producing antibodies in response. In contrast, passive immunity involves the acquisition of such antibodies from another animal or by administration. Maternal transfer of antibodies to offspring is an example of passive immunity. Behring's serotherapy is another example, as here the patients acquire antibodies passively by injection. Thus, in all such cases there is no active engagement of the host's immune system. The baby just sits contently sucking in the mother's antibodies from breast milk.

In 1890, Robert Koch invited Ehrlich to join his team at the newly established Institute for Infectious Diseases in Berlin (now the Robert Koch Institute). At that time, Emil von Behring was running into problems with his serotherapy. The responses to the treatment were variable: some patients were doing well, while in others the serum did not work as effectively as expected. Behring was instructed by Koch to approach another physician–scientist, one who, he said, might be able to discern the cause of these discrepancies. That man was Paul Ehrlich.

Ehrlich grudgingly agreed to help Behring – he preferred to work alone – and they worked together for six months, enriching the antiserum and standardizing the preparation. Finally, they found a serum of consistent strength that was consistently effective. Ehrlich had determined the amount of antitoxin required to neutralize a given amount of toxin. More a chemist than a physician, he considered the toxin–antitoxin reaction to be merely a chemical one, and began establishing the standard by which the amount of toxin in a serum could be precisely measured.

He also showed, like Glenny had done in London, that the toxins – especially those of diphtheria and tetanus – would, if left standing, heated or treated by various chemicals, change and become non-toxic. But they still retained their ability to make antibodies. These changes thus turned the toxin into a toxoid, no longer toxic but still capable of antibody formation. These toxoids could therefore be used as safer vaccines.

Ehrlich believed (incorrectly, as it happens) that only toxins could stimulate the formation of antibodies. The toxin, he suggested, 'can act firstly as a poison and secondly stimulate the production of specific antitoxin in the animal body'.[2] His study into the toxin–antitoxin interaction led to the formulation of his famous side-chain theory. According to Ehrlich, the tetanus toxin united with certain chemical receptors (poison receptors) causing disease.[3]

The part of the toxin that bound its receptor – he called it the 'haptophore' – was the same that had bound the antitoxin. That was why the antitoxin antibody was able to block the toxin from binding to its receptor, thereby neutralizing it. Ehrlich thought that these receptors were in fact identical to antibody molecules, and he called them 'side chains'. Although his theory is not completely accurate, Ehrlich's genius lay in the fact that he was able to appreciate that the three-dimensional structures of the antigen and antibody were complementary, allowing the specific binding of antigen to the antibody, like a key fitting a lock. This formed the basis of *specificity* of antibodies. 'As the cell receptor is obviously pre-formed,' Ehrlich wrote, 'and the artificially produced antitoxin only the consequence, i.e. secondary, one can hardly fail to assume that the antitoxin is *nothing else* but *discharged* components of the cell, namely receptors discharged in excess.'[4]

Ehrlich proposed that the antigen selected a side chain from the many available and bound to it specifically. This binding caused the cell to produce more side chains (antibodies) that would then be exported into the blood. Figure 6 depicts this theory in a diagram that Ehrlich used in his lecture to the Royal Society in London in 1900. Here we see antibody-producing cells with various toxin receptors (1). When a particular toxin binds its specific receptor (2), the cell makes more of the same type of receptor, which binds more toxin (3). Finally, the cell exports copious amounts of this receptor – the antitoxin antibody – that neutralizes the toxin (4). This was the earliest of a set of theories that sought to explain how antibodies were formed. We will come across these theories and how the mystery was finally resolved in Chapter 24.

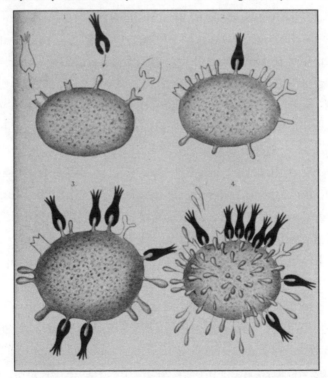

Figure 6. Ehrlich's side-chain theory. From P. Ehrlich, 'Croonian lecture', Proceedings of the Royal Society of London 1900; 66: 424.

Paul Ehrlich also carried out a series of experiments which demonstrated another important but curious feature of antibody immunity:

cross-reactivity. When he injected red blood cells from oxen into rabbits, the rabbits produced antibodies against the antigens found on the ox red cells. But to his surprise, when he mixed these rabbit anti-ox antibodies with red blood cells from goats, the anti-ox antibodies bound to the goat cells and caused them to disintegrate, a process known as 'lysis'.

What was going on here? It so happens that the antigens on both ox and goat red cells are structurally very similar, so rabbit antibodies raised against ox antigens could bind to goat red cells. So these antibodies were cross-reacting. Specificity, Ehrlich deduced, is not absolute. We will find out why and how the red cells exploded when the antibodies bound them in Chapter 13.

In his numerous experiments on haemolysis, Ehrlich found that, even though animals easily made antibodies against red blood cells from other animals of the same species (isoantibodies), they could not form antibodies against their own red blood cells. So, it appeared that the immune system was incapable of, or was somehow prevented from, acting against its own antigens. This he called *horror autotoxicus*.[5] Autoimmunity couldn't happen. In this assertion, we now know that he was plain wrong.

Ehrlich went on to make significant discoveries in immunology, histology and chemotherapy. He discovered Salvarsan, the first effective drug against syphilis. He was an indefatigable worker. Spending long hours in his small laboratory, he often skipped meals and chain-smoked around 25 strong cigars a day. He was seen scurrying through the long corridors of the institute, lost in thought, carrying a box of cigars under one arm and his pet dog under the other. But in his laboratory he was the undisputed master, whose authority could not be challenged. He sent out daily orders, which he called 'blocks', to all his staff, irrespective of seniority, which gave precise, detailed instructions on the tasks they had to carry out on that day.

Ehrlich was a leading humoralist, believing in the primacy of the antibody response in immunity. Unsurprisingly he often clashed with the leader of the opposing camp, the formidable Metchnikoff, who propounded his cellular theory of immunity. They continued this battle for decades until later findings eventually resolved the issue. Privately, Ehrlich and Metchnikoff were friends, and Ehrlich wrote warmly about a visit to Paris with his wife where they were guests of the Metchnikoffs. The scientific community, however, acknowledged the importance of both humoral and cellular factors in immunity, and in 1908 Paul Ehrlich and Élie Metchnikoff were jointly awarded the Nobel Prize in Medicine.

i   We now know that antibodies of the immunoglobulin G (IgG) class can cross the placenta and provide passive immune protection to the foetus. In fact, pregnant women are encouraged to get vaccinated during pregnancy to enable protective antibodies to get into the foetus from the mother. After birth the newborn is protected by these maternal antibodies for about six months, after which the baby's immune system begins to deal with any pathogens it may encounter, with adult levels of IgG being reached by the first year. Another way in which the newborn is afforded protection, albeit passively, is by antibodies that are found in colostrum. So when a baby suckles, these protective antibodies of the IgA class can provide protection to the baby, especially preventing gastrointestinal infection by the antibodies binding to and blocking attachment of pathogens to the gastrointestinal tract.

Chapter 12

THE DEADLY KISS OF THE ANTIBODY

The Institute of Infectious Diseases, Berlin, 1894. A young doctor sat hunched over his microscope. Outside in the corridor he could hear a loud commotion, but the former military doctor turned bacteriologist couldn't tear himself away from what he was seeing through the lens: something truly remarkable was unfolding, something never seen before.

Richard Pfeiffer, born in 1858, completed his medical studies at the Kaiser-Wilhelms-Akademie in Berlin. After graduating, he joined the army as a medical officer. But his passion was bacteriology, and he soon began to spend more time in the laboratory and less time on the hospital wards. In 1887, Pfeiffer began work at the Institute of Hygiene under Robert Koch.

At that time antibodies were considered to have just one function: neutralizing bacterial toxins. Pfeiffer made a remarkable finding that demonstrated a function of antibodies that was unrelated to their antitoxin properties, and this observation would be named after the man who discovered it.

Pfeiffer had worked his way up the ranks to become director of the Science Department at Robert Koch's institute in 1891. A cholera epidemic struck Hamburg in 1892 and Pfeiffer was assigned to study the disease. In one experiment involving the immunization of mice with two morphologically similar bacteria, V. *cholera* and V. *Metchnikoff*, Pfeiffer found that immunization with one cholera strain, while conferring immunity to that strain, did not protect against infection by the other strain.

But the key experiment that propelled Richard Pfeiffer to fame involved the demonstration that bacteria could be destroyed – exploded – by antibodies. Pfeiffer immunized guinea pigs with killed cholera bacteria and then injected them with the live vibrio bacterium. He then injected some saline into the abdominal cavity of the animals, gave the abdomen a squeeze to mix the saline and drew out a sample of the fluid containing the injected live bacteria. Placing a few drops of this fluid, called peritoneal fluid, from the animal on a slide and looking through the microscope, Pfeiffer observed the rapid destruction of the bacteria! Pfeiffer describes what he observed:

> Immediately after the injection, all the vibrios stopped moving. After 10 minutes, many granules and swollen vibrios but almost no leucocytes were seen. After 20 minutes of the infection, all the vibrios disappeared and many granules remained. Approximately 95% of the granules were extracellular and 5% were in the protoplasm of the leucocytes. Before my eyes, the cholera vibrios were lysed free of phagocytic influence… I obtained the same results with a passively immunized guinea pig. The animal body plays a major and active role. It reacts to the stimulus of the vibrios under the influence of the immune substances in serum to produce the bactericidal activity.[1]

The crucial point he made was that the destruction was 'free of phagocytic influence'. So antibodies generated by immunization were most likely reacting with the live cholera bacteria introduced into the peritoneal cavity of these animals, causing their lysis. This bacterial lysis by antibodies, discovered in 1894, was called the 'Pfeiffer Phenomenon'. Antibodies were not only neutralizing toxins: they could also cause bacteria to explode!

Pfeiffer then found that he could transfer this property between animals. He showed this by first taking peritoneal fluid from an immunized guinea pig and then injecting this fluid into an unimmunized guinea pig. When this 'normal' guinea pig was then injected with live cholera bacteria, the antibodies from the peritoneal fluid of the previously immunized guinea pig were able to lyse the bacteria (Fig. 7).[2]

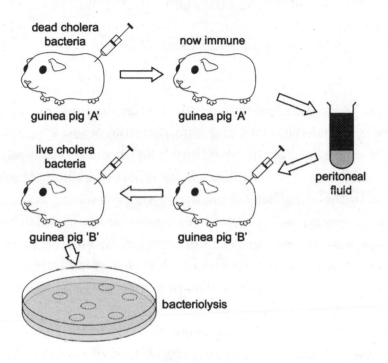

Figure 7. Transfer of the Pfeiffer Phenomenon.

So here we come across another important function of antibodies: they can bind to antigens on bacteria and cause their destruction. But how? One method, as we shall see in Chapter 13, is by activating a series of serum proteins collectively called 'complement', which leads to holes being punched in the bacterial cell wall, causing their demise. The other method is by allowing macrophages to bind to the antibody that has already bound to the bacteria: this makes it easier for macrophages to ingest and phagocytose bacteria. It is rather like sugar-coating a doughnut to make it more enjoyable and more likely to be eaten. We shall now explore these important mechanisms in more detail.

Chapter 13

LETHAL ASSISTANTS

Paris, 1894. The Pasteur Institute. A small, thin, nondescript man with pale-blue eyes and a thin, curling moustache stood in front of the head of the laboratory. The laboratory head was Élie Metchnikoff, and he began reading the report which he had just been given by the young Belgian physician who was staring at him with strange, dreamy eyes the colour of a lake in springtime. Metchnikoff ran his hand through his own unkempt, greasy hair. He didn't like what he was reading. He realized with a sense of frustration, tinged with admiration, that the softly spoken Belgian had stumbled on something significant.

The pendulum swung back in the direction of the French school when the Pfeiffer Phenomenon caught the eye of the Belgian physician who had begun to work at the Pasteur Institute. Jules Bordet was born in Soignies in 1870 and received his doctorate from the Free University of Brussels in 1892. In 1894, his talents in research led to a scholarship that enabled him to leave Belgium to work at the prestigious Pasteur Institute. Bordet, a quiet, unassuming but methodical worker, was, unfortunately for him, assigned to work under the unpredictable and often volatile Metchnikoff. Carrying out his research surreptitiously, for any work that was not directly related to cellular immunity would unleash the wrath of his Russian mentor, he began exploring the Pfeiffer Phenomenon.

In 1895, the 25-year-old Belgian carefully repeated Pfeiffer's experiment and found that immune serum from guinea pigs previously immunized with the cholera bacterium, when introduced into a normal healthy guinea pig and subsequently challenged with cholera bacteria, caused bacterial lysis observable *in vitro*. But, crucially, Bordet discovered that, if the immune serum was old or had been heated beyond 55°C, there was no bacterial lysis. The lysis could be restored, however, by adding fresh serum from an unimmunized healthy guinea pig.

Realizing that this was a significant finding, Jules Bordet concluded that to obtain bacterial lysis one needed two things: a specific anti-body that appeared to be temperature-stable and a heat-sensitive component found in all animals, irrespective of whether they were sensitized or not. He called this heat-sensitive component 'Alexin'.[1] It was later called 'complement' by Ehrlich, because it appeared to *complement* antibodies.[i]

Three years after the discovery of complement, Jules Bordet found, as Ehrlich had shown previously, that red blood cells taken from one animal and injected into another were also lysed, but this reaction required the presence of complement. In 1901, he developed the complement fixation test, a procedure that was useful for the detec-tion of specific antibodies.[ii] This was the method subsequently used by August Paul von Wassermann in his syphilis test.[2]

At first the findings of Richard Pfeiffer and Jules Bordet greatly distressed Metchnikoff, who fought off these challenges to his cel-lular theory with vigour. And to make matters worse Bordet was one of Metchnikoff's students! But eventually, albeit grudgingly, he began acknowledging their significance. For the discovery of comple-ment, Jules Bordet was awarded the Nobel Prize in and Physiology or Medicine in 1919.

But there was still some hope for Metchnikoff, and ironically it came from Germany, from Robert Koch's work on tuberculosis. It appeared that there were instances when immunity could *not* be demonstrated to be mediated merely by toxin-neutralizing, bacteria-popping antibodies found in blood. Immunity to tuberculosis was such an example. Circulating levels of anti-tubercle antibody were not correlated with immunity. Immunity to the disease could not be transferred by serum antibody. What was going on here? Was there some other factor or factors conferring immunity to tuberculosis? But, as Koch had demonstrated by his tuberculin studies, it was probable that cellular factors were responsible for immunity to tuberculosis.

Notwithstanding these findings, the war between the cellularists and the humorists continued unabated. It would take an eccentric Englishman to bring the two disparate theories together finally, and to demonstrate the importance of *both* cellular *and* humoral factors in immunity. His name was Almroth Wright.

*

Born in Yorkshire in 1861 to an Anglo-Irish father and a Swedish mother, Almroth Wright obtained his medical degree in Dublin in 1883. In the late 1890s he found himself working as a civilian scientist for the British army. A successful research career led to his appointment as professor of pathology at Saint Mary's Hospital, London. Working at the Army Medical School in Netley, Wright became interested in exploring Pasteur's methods for developing a vaccine against typhoid. Wright prepared a vaccine using an attenuated strain of typhoid bacillus, and tested it on his students, who, after developing some rather uncomfortable symptoms, appeared to acquire resistance against typhoid. To reduce the annoying side effects, Wright began using a safer, killed typhoid vaccine, testing it first on himself and his colleagues before recommending it to the army, by whom it was eventually adopted.[3]

Wright's cholera vaccine, like Jenner's cowpox and Pasteur's anthrax vaccines, was a prophylactic, one meant to prevent people from getting the illness in the first place. He also explored the idea of therapeutic vaccination, whereby a vaccine was given to individuals who already had the infection. This was the approach that Pasteur had used with his rabies vaccine and Koch had attempted with his tuberculin therapy. Both approaches were examples of active immunization, whereby the host produced an immune response to the vaccine. It is worth repeating that this was very different from the approach of passive immunization, where pre-formed antibodies were given to the host to combat an infection that was well under way – the approach used by Behring in his serum therapy. Ehrlich's guinea-pig suckling experiments, as we saw earlier, were also examples of passive immunity.

Almroth Wright's vaccines, while working well for some, were ineffective in others. He decided to consider the problem of individual differences in the immune responses against his vaccines: some patients showed excellent immunity whereas others showed a weak or ineffective response. Wright wanted to know why.

He devised a simple test that, he believed, would indicate the efficacy of the immune response to a given pathogen, one that would be useful in predicting whether a vaccine would work for a particular person or not. Wright believed that the test would allow him to adjust the dose of the vaccine to suit the different degrees of immunity shown by different individuals. So someone who was a weak responder may require a higher dose of a vaccine to obtain an effective response. Wright decided to use Metchnikoff's phagocyte as the indicator of immunity. Perhaps, he told his students, the process of phagocytosis could be quantified to indicate the immune status of an individual. However, Wright found that if he separated the phagocytes from blood, the cells carried out phagocytosis only weakly. When the cells were mixed with blood, the macrophages sprung back into action. Blood appeared to have something that enhanced phagocytosis: some component that prepared the microorganisms for ingestion and subsequent destruction within the phagocyte. He called these substances 'opsonins', from the Greek word *opsōnéō*, meaning 'to purchase provisions'.[iii]

Peering down his microscope, Wright observed macrophages attempting to digest staphylococcus bacteria that he had isolated from a boil. He counted the numbers of bacteria that surrounded and adhered to these phagocytes. He found that this process was greatly enhanced when fresh serum was added – that is, serum containing his opsonins. It is now understood that Wright's opsonins are antibody and complement, as first described by Jules Bordet. Antibodies and certain complement proteins would bind to and coat the bacteria and prepare them for phagocytosis. It was later found that phagocytes, like macrophages, had specific receptors for antibody and complement, which allowed them to bind these factors that were coating the bacteria. This greatly enhanced their phagocytic ability. But at the time Wright had no idea what these opsonins were. He simply knew that they enhanced phagocytosis. When he diluted the serum, he found a corresponding decrease in phagocytosis, indicating that the opsonins in the serum had been diluted.

Wright's test involved a formula, the 'opsonic index', which was the number of microorganisms attached to white blood cells when the cells were suspended in a serum sample divided by the number of microorganisms attached to white blood cells in control serum. Therefore, the opsonic index gave an indication of the quantity of opsonins in a person's blood and provided a measure of the individual's ability to fight off an infection by that particular organism. He used the index to assess the ability of a patient to resist infections caused by staphylococci, such as boils and abscesses. Wright also used the opsonic index to assess the ability of a patient to respond to various therapeutic vaccines. If a patient suffering from a staphylococcal infection showed a high opsonic index, this would indicate a brisk immune response and a probable efficacious response to a vaccine. If, on the other hand, the patient showed a low opsonic index, they would possibly require a higher dose of the vaccine to elicit an effective immune response.[4]

But in practice, using the opsonic index to predict vaccine responses proved to be inaccurate. The test also did not work on certain infections, notably pulmonary tuberculosis. However, Wright had demonstrated something important. He had shown that the cellular and the humoral arms of the immune responses worked together in the fight against pathogens. Phagocytes worked best when opsonins from serum were present. He documented his findings in a book that he dedicated to Élie Metchnikoff and Paul Ehrlich, both of whom he greatly admired.[5]

Writing about Wright's contribution to immunology, John Turk mentions an interesting meeting between Almroth Wright and the playwright George Bernard Shaw.

In 1905 George Bernard Shaw visited Wright in his laboratory at St Mary's to see his work on phagocytosis and the assessment of the patient's resistance to infection by the opsonic test. As a result, he made Almroth Wright his model for Sir Colenso Ridgeon in his play *The Doctor's Dilemma*, first produced in 1906. In this play Shaw is well aware of the opsonic theory of phagocytosis:

'What it comes to in practice is this. The phagocytes won't eat the microbes unless the microbes are nicely buttered for them. Well, the patient manufactures the butter for himself all right; but my discovery is that the manufacture of that butter, which I call opsonin, goes on in the system by ups and downs – Nature being always rhythmical, you know – and that what the inoculation does is to stimulate the ups or downs, as the case may be... Drugs can only repress symptoms; they cannot eradicate disease. The true remedy for all diseases is Nature's remedy. Nature and Science are at one, Sir Patrick, believe me; though you were taught differently. Nature has provided, in the white corpuscles as you call them, in the phagocytes as we call them – a natural means of devouring and destroying all disease germs. There is at bottom only one genuinely scientific treatment for all disease, and that is to stimulate the phagocytes. Stimulate the phagocytes.

Drugs are a delusion. Find the germ of the disease; prepare from it a suitable anti-toxin; inject it three times a day quarter of an hour before meals; and what is the result? The phagocytes are stimulated; they devour the disease; and the patient recovers – unless, of course, he's too far gone. That, I take it, is the essence of Ridgeon's discovery.'[6]

But with the emergence of a novel therapy for infectious diseases, Almroth Wright's therapeutic vaccination strategies gradually fell out of favour. A chance observation made by one of Wright's students working at St. Mary's Hospital led to a revolutionary new way of treating infectious diseases. His name was Alexander Fleming, and he discovered the first chemotherapeutic agent, penicillin. But, as we shall see later, therapeutic vaccination strategies may still have a role to play in our constant battle against pathogens.

i   When we talk about 'complement' we are referring to a group of around 30 different serum proteins and glycoproteins. These complement proteins are mainly made in the liver and exported into the blood. In the blood these proteins circulate as functionally inactive forms – many as proenzymes with a masked active site. The active sites are then exposed by cleaving – cutting off a bit of the molecule – by other proteins. These activated compounds have short half-lives as they would be too dangerous to be left for long in the bloodstream. The active proteins are components of several pathways, or cascades, that help to get rid of pathogens by several mechanisms. Some of these complement proteins generate the inflammatory response; others join together to make a plug that gets inserted into the cell wall of the bacterium, causing it to lyse.

Other proteins of the complement cascade, opsonins, coat pathogens, making them bind to macrophages that have receptors for the complement proteins. All these mechanisms – lysis, opsonization and generating the inflammatory response – help the immune system destroy invading pathogens.

Another important function of some complement proteins involves binding to antigen–antibody complexes. Antigen–antibody complexes can be formed as a consequence of immune responses against pathogens. Again, if these complexes were left lying around they could cause problems: for example, immune complexes can get lodged in kidney glomeruli and lead to inflammation. Therefore, an important function of complement proteins is to degrade immune complexes. They do this by binding to antigen–antibody complexes and then taking these to the liver and spleen, where the complexes are safely destroyed, thus preventing immune complex disease.

ii Adding complement to red blood cells coated with anti-red blood cell antibodies would cause lysis of the cells, by the insertion of a plug into the cell membrane. Thus, coated red cells would constitute an 'indicator system' that can be used to determine the presence of a specific antibody against a given antigen. So, in the complement fixation test, a serum sample is mixed with antigen and added to the indicator system. If a specific antibody was present in the serum it would bind to antigen and fix or use up the complement. The complement, having been used up, will not be able to lyse the coated red cells (the indicator system).

iii We now know that the opsonins described by Almroth Wright are antibody and complement. One of the most important opsonins is C3b; this is a complement protein that coats bacteria and binds to CR1 receptors on macrophages, allowing the macrophage to phagocytose the bacteria with greater efficiency. The macrophages also have receptors for antibody*(FcR) and again this allows the macrophages to bind more effectively to antibody-coated bacteria.

Chapter 14

THE DREAM OF DR WILLIAM COLEY

New York, 1890. Elizabeth 'Bessie' Dashiell, a bright, energetic, 17-year-old socialite turned up at the New York Hospital after injuring her right hand while travelling in a train car. Even though the injury had happened a few weeks previously, the hand was still swollen and painful. She was seen by Dr William Coley, a young orthopaedic surgical intern. Coley noted a lump in her right hand and took a biopsy of the swelling, hoping to find inflammatory changes in line with the traumatic injury. Instead, he found an aggressive type of bone cancer, a sarcoma. Coley had no choice but to amputate Bessie's right arm just below the elbow. But it was too late, and Bessie died in January 1891, just ten weeks after diagnosis. Autopsy revealed that the cancer had spread throughout her body.

The death of his young patient greatly affected Coley, who determined to find an effective treatment for bone cancer; he would spend the rest of his career chasing this dream. Coley started off by reading the case histories of all patients who had been treated for sarcoma at the New York Hospital. He was looking for a clue, any clue, that might point him in the direction of a possible cure. Eventually he came across a case that caught his attention.

The patient was a 31-year-old German immigrant called Fred Stein, who had presented with a round cell sarcoma of the neck. The tumour had been removed, but had recurred. This, Coley read, had happened on five occasions. After the fifth excision the surgeons had given up, deeming the case hopeless. Stein had then developed an infection characterized by a spreading rash and high fever: erysipelas, caused by *Streptococcus pyogenes*, the bacterium also responsible for the common sore throat. But then something strange had happened. During the erysipelas infection Stein's tumour had regressed and disappeared. The tumour had, however, reappeared once the infection had resolved, only to disappear again when the infection had made a second appearance some two weeks later.

This had all happened seven years ago, and Coley was keen to find out what had happened to Stein. The hospital records showed that he came from Lower Manhattan, and so Coley began searching for Stein, combing through the rundown slums of that part of the city. After several weeks he found the patient. Seven years since he had been discharged from hospital Stein was in good health, with no evidence of the tumour that had been deemed inoperable.

Coley plunged into the literature, looking for similar reports of tumour regression after a patient had contracted an infection. He found several. In 1725, the French mathematician and physician Antoine Deidier had observed that patients who had the misfortune to contract syphilis did not seem to suffer from malignant tumours. In 1867, the German physician Wilhelm Busch had described the disappearance of a tumour when one of his patients developed erysipelas. In 1882, Friedrich Fehleisen had observed the vanishing of a tumour when he injected a patient with streptococci, inducing erysipelas.[1] In all, Coley found 47 cases where various infections had caused regression or complete disappearance of tumours. Of 38 cases where erysipelas infection had been concurrent with tumour, 12 showed the complete disappearance of the tumour, while the rest showed a reduction in its growth.[2]

In 1891, William Coley began treating sarcoma patients with streptococcus. Signor Zola, a 35-year-old Italian migrant, had presented with a recurrent sarcoma of the neck which was also found in the tonsils. The tumour, the size of a hen's egg, was deemed inoperable. The mass almost totally blocked Zola's throat, and the patient was regurgitating liquids through his nose. Zola looked in bad shape: he was weak and cachectic, and was only expected to live for a few weeks. William Coley decided to inject him with a culture of *Streptococcus pyogenes* every three to four days for several months. Erysipelas did not develop, even though Zola's tumour shrank, and his overall condition improved. Whenever Coley stalled the injections, the tumour started to grow. Coley wanted to try a more virulent culture, and this he obtained from the laboratory of Koch in Berlin. He injected some of Koch's streptococci directly into the tumour. Within the hour, Zola developed a high fever with chills. He was also in pain and began vomiting. After 12 hours the typical erysipelas rash covered the tumour, extending onto his face and head. The symptoms continued unabated for over a week, then, much to Coley's delight, the tumour started to soften, began to shrink and disappeared after two weeks. The tonsil tumour, however, didn't disappear completely, and became a hard, fibrous mass. Zola was able to eat and drink and slowly made a full recovery. He lived for another eight years before the tumour returned and the Italian finally succumbed to his illness.[3]

In his early trials Coley treated ten patients suffering from inoperable cancers – carcinoma and sarcoma – with live streptococcus. Even though he couldn't induce erysipelas in six patients, their tumours got smaller: they showed partial regression. In the four patients in whom Coley managed to induce erysipelas, the tumours went into full remission. Coley soon realized that to induce full remission of the tumours, the patients must develop erysipelas. He quickly found that this was not always possible and that there were also instances when the erysipelas infection itself caused the death of the patient.[4]

Coley tried using cultures killed by heat sterilization, but these preparations had negligible effect. He then tried a mixed culture, which came to be called 'Coley's toxin', made up of heat-killed streptococci and another bacterium, *Serratia marcescens*. The first patient to receive Coley's toxin was a 16-year-old German boy with an abdominal sarcoma attached to the pelvis and judged to be inoperable. It had also infiltrated the bladder. The tumour responded to Coley's toxin and regressed, finally disappearing a few months later. The man lived for another 26 years before dying of heart disease.

By 1916, Coley had tried the toxin on around 90 cases, mainly using it on patients suffering from inoperable sarcomas and finding the method to be less effective against other types of cancer. In 1899, Coley's toxin was produced commercially and was widely used for the next 30 years. It has been estimated that by the time he retired Coley had treated around a thousand patients in this way.

Even though Coley's toxin treatment was clearly successful in many cases, the doctor wasn't beyond criticism. Coley didn't always follow up on his patients, and his methods were inconsistent. He used some 13 different preparations of his toxin, some of which were clearly more effective than others. The mode of administration also differed: some were given the toxin by injection into a vein, others by injection under the skin or into a large muscle, while in some cases the toxin preparation was directly injected into the tumour mass. Hence, unsurprisingly, when other doctors tried to reproduce Coley's method they had inconsistent results. Eventually Coley's radical method of immunotherapy fell out of favour. The refusal of the US Food and Drug Administration (FDA) in 1962 to acknowledge Coley's toxin as an approved drug was the death knell. Another reason for the demise of Coley's toxin was timing, as radiation treatment for cancer was developed and gave more consistent and reproducible results. Chemotherapy against cancer, which was introduced in the 1940s, also led to Coley's toxin being sidelined and consigned to medical history.

But Coley's dream didn't end there. His cause was taken up by his daughter, Helen Coley Nauts, who, according to Edward McCarthy's review of Coley's work, 'devoted her life to the study of her father's toxins. She tabulated every patient he treated – over a thousand cases – and reviewed all his notes. She published 18 monographs and noted that in 500 of the cases there was near-complete regression.'[5] Lloyd Old, associate director of the Memorial Sloan Kettering Cancer Center, said of Coley's work: 'Those who have scrutinized Coley's results have little doubt that these bacterial toxins were highly effective in some cases.'[6]

A review of 224 cases of spontaneous tumour regression conducted in 1971 found that in 62 of such cases there was an active infection accompanied by fever. In another study, looking at metastatic melanoma cases, 21 out of 68 spontaneous remissions occurred when there was a concurrent infection, and in 11 cases regression followed some form of immunological treatment such as Bacillus Calmette–Guérin (BCG) vaccination, antibody treatment or injection of tumour cells.[7]

A few clinical trials using Coley's toxin have been carried out more recently, but they have not shown consistent results. But there have been some dramatic findings. A phase-one study carried out in 2007 to examine the safety and effectiveness of the toxin included a patient with metastatic bladder cancer who, upon treatment, showed a clear response, achieving a 50 per cent reduction in his tumour burden.[8]

So how did Coley's toxin work? How did the infection and the fever lead to the regression or disappearance of the tumour? The short answer is we don't know for sure.[i] Until the late 1980s, many did not believe that the immune system played a key role in controlling cancer. After all, cancer cells were our own cells, and even though they had turned malignant they were still technically 'self'. In his 2003 review of the role of the immune system in cancer, Christopher Parish wrote: 'I distinctly recall that by 1980 cancer immunotherapy was generally regarded as an approach with little or no chance of success.'[ii9]

Nevertheless, it is interesting to note that, back in 1909, Paul Ehrlich had postulated that the immune system may be involved in the control of cancer. But there was limited evidence to support his contention at the time, and even that was questionable. However, one modern application of immunotherapy against cancer has vindicated William Coley's tireless efforts: immunization by the BCG vaccine is regarded as the one of the most effective ways to treat superficial bladder cancer.

i   The gram-negative endotoxins from *Serratia* in Coley's toxin, by binding to toll-like receptors (discovered in the 1990s) on a variety of immune cells such as macrophages and dendritic cells, generate signals within these cells, which lead to several effector mechanisms that mediate and amplify immune responses against tumours. These include transcription of cytokine genes, causing the production of anti-tumour cytokines such as interleukin 2, interferon alpha and tumour necrosis factor. The cytokines also activate resting dendritic cells, which, by presenting tumour antigen to T cells, generate T-cell responses against the tumour.

Fever, a component of the innate immune system, has also been shown to exhibit anti-tumour effects. Fever can activate heat-shock proteins, which in turn can activate immature dendritic cells, which will then present tumour antigens to T cells, leading to responses against the tumour mass. Elevated body temperature has also been shown to cause tumour cell apoptosis directly. Hence Coley's toxins could, through cytokines, activate both the innate and adaptive immune systems against cancer.

ii In the 1950s, researchers began transplanting tumours into animals. Most of these transplanted tumours were rejected by the host animals, but since most of these experiments involved outbred animals this was put down to immune responses against foreign antigens on the tumours. Then, in the late 1950s, evidence began to trickle through to suggest that the immune system may have a role against cancer. When a tumour was removed from an animal and reinjected, it was rejected, suggesting that a secondary response had caused the rejection. This led to speculation that tumour-specific or tumour-associated antigens were being recognized by the immune system. This led to the idea of immunosurveillance: immune lymphocytes were patrolling the body, eliminating any tumour cells they came across. When in the 1980s and 1990s tumour-specific antigens (for example, HPV, E6 and E7 proteins induced by viruses) and tumour-associated antigens (such as CEA, PSA and AFP) were being discovered, the idea of immune responses against tumours began to be considered seriously. Interestingly, the tumour-associated antigens were found to be 'normal antigens' that were expressed either only during foetal life or in adults at higher than normal levels.

According to Parish, 'the single most powerful piece of evidence in favour of tumour-specific immunity was the demonstration that highly malignant cells are profoundly genetically unstable... and because of this genomic instability large numbers of novel tumour-associated antigens are generated that have never been seen by the host's immune system'.[10] Additionally, immunodeficient mice were shown to exhibit much higher rates of cancer.

Chapter 15

THE BLOOD OF STRANGERS

New York, 1 June 1943. At the Rockefeller Institute, a man was found slumped over his laboratory bench. The technician who discovered him immediately recognized him. Thin, well over 70, his sparse hair wispy and grey, he appeared to be barely alive, but was holding a pipette in his right hand. He had suffered a heart attack. He was rushed to hospital but died two months later. Few at the institute really knew him, for the man was a loner who shunned the company of others, preferring to live quietly and unobtrusively, especially after the death of his wife some sixth months previously. But everyone at the Rockefeller Institute knew of him, for he was a true giant of medicine: a Nobel laureate, the Jewish haematologist who had escaped the Nazis by moving to America from his native Vienna before the outbreak of the Second World War. He had almost single-handedly revolutionized the science of immunology and haematology. His seminal discoveries would save the lives of millions, for his efforts had paved the way for safe blood transfusion. His name was Karl Landsteiner, and he had discovered the ABO blood group system in humans.

Landsteiner was born in Vienna in 1868 and graduated from Vienna Medical School in 1891. After the First World War, Vienna was in economic decline and Landsteiner decided to emigrate. First, he moved to the Netherlands and worked as a prosector in the Catholic St Joannes de Deo hospital in The Hague. But the pay was appalling and, struggling financially, he had to supplement his income by working in a small factory that produced old tuberculin (*Tuberculinum prestinum*). Finally, Landsteiner accepted an invitation to work at the Rockefeller Institute in New York, and he moved to the United States in 1923.

The English physician William Harvey had discovered the circulation of the blood in 1628, and since then many investigators had explored the possibility of blood transfusion. Although it would have been clearly desirable if animal blood could be successfully transfused into humans, it soon became apparent that this would not work. Inter-species transfusions always resulted in the rapid lysis of the transfused red cells: in 1875, Leonard Landois, a German physiologist, showed that when red blood cells from animals were transfused into humans they clumped and lysed. It was subsequently shown that lysis of the red cells was mediated by a factor in the serum: when red blood cells from one species were mixed with serum from a different species they lysed.

Landsteiner suggested that a similar process of destruction of donor cells could also occur when blood from one individual was transfused into another. His ideas, however, did not meet with much enthusiasm until he carried out a series of groundbreaking experiments.

Landsteiner found that some individuals contained naturally occurring antibodies that reacted with certain antigens found on the red blood cells of others. In 1901, he carried out an experiment looking for these natural antibodies in six individuals, including himself. He randomly mixed red cells and sera from the six individuals and observed the pattern of clumping or agglutination. Based on the ability of a given serum to agglutinate red blood cells from different individuals, Landsteiner was able to place individuals into three distinct groups. This work led to the formulation of 'Landsteiner's Rule'. Serum from an individual contained antibodies against antigens that were *not* present on that individual's red blood cells. This simple experiment eventually led to the characterization of the ABO blood system and to successful blood transfusions. Initially he classified individuals into three groups based on the antigens found on their red cells: groups A, B and O. A fourth group, AB, was described by his students in 1902.[1]

For example, a group-A person whose red cells contain the A antigen will have anti-B antibodies in their blood, whereas a group-B person with B antigen on their red cells would have anti-A antibodies. Clearly, someone with group-AB red cells, having both the A and the B antigen, will not have any anti-A or anti-B antibodies in their blood (for they would immediately bind and destroy their own red cells!). The fourth group is O, where the red cells do not have A or B antigen, but there are anti-A and anti-B antibodies in the blood (Fig. 8). Because they don't have A or B antigens, red blood cells from group-O individuals will not be lysed by natural antibodies. Group-AB individuals, on the other hand, don't have anti-A or anti-B antibodies that would bind to A or B antigens on red blood cells. Other minor-blood-group antigens have been subsequently discovered, but the ABO system is the main one that is relevant to blood transfusion.

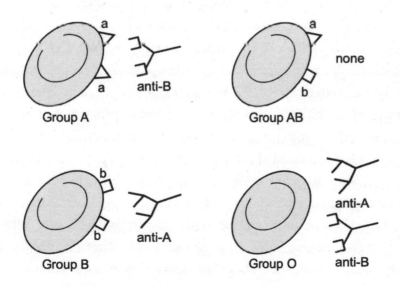

Figure 8. Landsteiner's Rule: ABO blood system in humans.

Landsteiner had a formidable opponent: Paul Ehrlich. As we know, Ehrlich had worked on intra-species haemolysis, and he strongly disagreed with Landsteiner's findings. Speaking at the seventy-third annual meeting of German scientists and doctors, he dismissed the agglutinating natural antibodies as mere by-products without biological relevance. 'What good is it to a goat,' he asked, 'if it has antibodies within its blood directed against the red blood cells of another animal?'[2]

In 1923, Landsteiner discovered haptens. These small chemicals were not immunogenic by themselves, but when coupled with a carrier protein provided antigenic determinants against which antibodies could be formed. Therefore, haptens were antigenic but not immunogenic. He studied the specificity of antibodies by observing whether one anti-hapten antibody could bind to a slightly different hapten, thereby indicating cross-reactivity. His work helped gain a clearer understanding of the specificity between antigens and antibodies.

Landsteiner also investigated Koch's tuberculin reaction. Using the hapten picryl chloride, he showed that the skin reaction, which was later called the delayed-type hypersensitivity reaction, could be transferred by the cell fraction and not by serum. Landsteiner showed this by first immunizing guinea pigs with picryl chloride and then transferring a cellular fraction from these animals into unimmunized guinea pigs that were subsequently challenged with the hapten applied to the skin. The unimmunized or naive guinea pigs developed the typical skin reaction some 24 hours later.[3] As they had not come across picryl chloride before, the animals must have produced the skin reaction – caused by the heightened secondary response – from the transferred cells. When this experiment was repeated with the serum fraction, there was no skin reaction. Also, if these transferred cells were killed prior to the transfer, there was no skin reaction. Landsteiner's colleague, Merrill Chase, repeated these experiments a few years later using *Mycobacterium tuberculosis* as the antigen, and got the same results.[4] So once again the pendulum swung, and some immune responses were shown to be clearly cellular and not due to antibodies. But it was for the discovery of human blood groups that Karl Landsteiner was awarded the 1930 Nobel Prize in Physiology or Medicine.

Chapter 16

FRIENDLY FIRE: AUTOIMMUNITY

In 1881 Paul Ehrlich came across a patient with a strange constellation of symptoms that appeared when he was exposed to cold temperatures. Within a few minutes to several hours of exposure to cold, the patient, who had a history of syphilis, developed back pain, cramps, chills and nausea and began passing dark-brown urine. The dark urine intrigued Ehrlich, who believed that haemolysis, the breakdown of red blood cells, was taking place within the patient when he was subjected to cold conditions. To test his theory, Ehrlich tied a ligature around one of the patient's fingers and immersed it in ice water. When he subsequently drew some blood from the finger he found that the red cells had haemolysed.

But what was causing this patient's red blood cells to pop? Ehrlich ruled out the possibility that this was an autoimmune phenomenon: he had never been able to find autoantibodies when an animal was given an injection of its own red blood cells. From this and other similar experiments involving blood injections between species, Ehrlich had devised his idea of *horror autotoxicus*: autoimmunity cannot happen. As for the syphilitic patient with the haemolytic finger, Ehrlich believed that a toxin was being released from blood-vessel walls at cold temperatures, and this was causing the haemolysis.[1]

*

In 1904, Karl Landsteiner and Julius Donath came across three patients who passed blood in their urine when exposed to cold. This disease was called 'paroxysmal cold haemoglobinuria', and, believing this to be an autoimmune reaction, they decided to carry out a series of experiments. First, they drew some blood from the patients and separated the plasma from the red cells. Then they mixed the separated plasma with red blood cells in a test tube and placed the mixture in ice water. When the mixture was subsequently placed in an incubator and warmed, haemolysis occurred. When the same thing was done with plasma and cells from a healthy person, haemolysis did not occur. The unusual haemolytic reaction appeared to be unique to the patients. Landsteiner and Donath then mixed a sample of patient plasma with red blood cells from a healthy individual (presumably with the same blood type) and repeated the experiment, placing the mixture in an ice-water bucket and then in a warm incubator. Again, haemolysis happened. They concluded that the patient's plasma contained the haemolytic agent, which, when exposed to the cold, bound the patient's red blood cells. Warming the mixture then caused haemolysis. In another, similar experiment, they found that if the patient's plasma was pre-warmed, haemolysis did not occur. However, when they added complement to this serum, haemolysis capacity was restored (Fig. 9).[2]

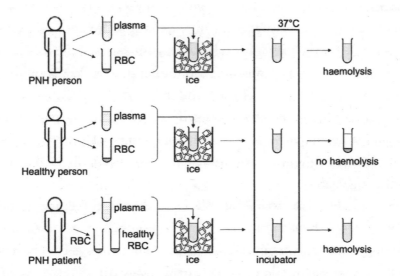

Figure 9. Cold haemolysis in paroxysmal nocturnal haemoglobinuria.

Yet another ingenious experiment followed. Landsteiner and Donath found that pre-absorbing the patient's antibodies onto red cells, thus separating them from serum, rendered the serum incapable of causing the cold-haemolytic reaction. After analysing the results of these experiments Donath and Landsteiner concluded that the cold-haemolytic reaction had two phases: absorption of patient autoantibodies to their own red blood cells at cold temperatures followed by lysis of these sensitized cells at warm temperatures if complement was present.[3]

But what caused these autoantibodies to be formed in the first place? In nineteenth-century Europe, most patients who had paroxysmal cold haemoglobinuria also suffered from congenital or adult syphilis. Landsteiner found that components in the sera from syphilitic patients reacted not only with organ extracts that contained the syphilis spirochetes but also with extracts made from uninfected tissues. He suggested that the autoantibodies in paroxysmal cold haemoglobinuria, formed as part of a normal antibody response against spirochete antigens, were somehow capable of binding to the patient's own red blood cells at cold temperatures. As paroxysmal nocturnal haemoglobinaemia was almost always observed in syphilitic patients, Landsteiner suggested that cross-reactive antibodies were responsible for the haemolysis, heralding the birth of the concept of molecular mimicry as a cause of autoimmunity. In molecular mimicry, a foreign antigen has structural similarity with a self-antigen, leading to an autoimmune attack on self-tissue. Currently, paroxysmal nocturnal haemoglobinuria is seen following viral infections, and the cold haemagglutinin antibodies are most likely directed against viral antigens.[4]

Not only had Landsteiner and Donath described the first autoimmune disease in humans but they had also proposed a mechanism for its aetiology. So autoimmunity could happen! But such was Paul Ehrlich's dominance at the time that few accepted Landsteiner's interpretation of his findings. *Horror autotoxicus*, it appeared, was here to stay.

*

But, slowly, more and more autoimmune phenomena began to be described. In 1946, James Gear, from South Africa, described a complication of malaria, blackwater fever, whereby red cells of malaria patients were apparently being lysed by autoantibodies.[i5] In the same year, the then veterinary student Robin Coombs developed a test that would be able to detect autoantibodies. The test was originally used to detect maternal serum antibody against the newly discovered rhesus antigen found on some foetal red blood cells. If there were such antibodies in the mother's blood sticking to the foetal red cells, the Coombs Test would detect them. But how? Coombs used an ingenious anti-antibody made in a rabbit to detect such anti-rhesus antibodies. The rabbit anti-human antibodies – made by immunizing a rabbit with human antibodies – are used to identify any anti-rhesus antibodies that may have bound onto rhesus antigen on red cells. When rabbit anti-human antibodies are mixed with rhesus positive cells that are coated with anti-rhesus antibodies, the rabbit anti-human antibodies, by binding to the anti-rhesus antibodies coating the cells, form a lattice that would cause the cells to clump or agglutinate. So, by detecting the presence of maternal anti-rhesus antibodies, the test could prevent haemolytic disease of the newborn caused by rhesus incompatibility (Fig. 10).[ii6]

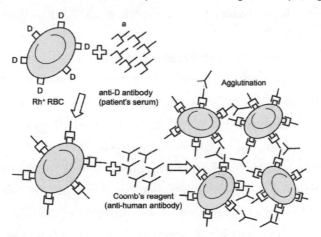

Figure 10. Detecting anti-rhesus antibodies using the Coombs Test.

Realizing that the test could be used to detect antibodies against a variety of antigens, Coombs applied his test to autoantibodies that caused a haemolytic anaemia in some patients. Thus, autoimmune haemolytic anaemia, whereby autoantibodies bound and lysed red blood cells, became the first autoimmune disease for which there was a specific test. In 1951, this type of immune haemolytic anaemia became the first disease to acquire the prefix of 'auto'.[7]

In 1951, William Harrington and others transferred serum from patients with idiopathic thrombocytopenic purpura, a disease in which the patient's platelets – cells in the blood responsible for clotting – were being destroyed by anti-platelet autoantibodies, into healthy volunteers (including Harrington himself) and noticed an immediate decrease in platelet numbers. A serum component, later found to be an autoantibody, was destroying the platelets. However, after a few days the platelet numbers were seen to recover.[8]

Subsequently, observations were made of human maternal autoantibodies causing disease in the newborn by entering the foetal circulation through the placenta, thus transferring the mother's disease to the child. This research demonstrated the autoimmune aetiology of several diseases such as thyrotoxicosis, myasthenia gravis and certain forms of systemic lupus erythematosus. The evidence that autoantibodies were responsible for the pathology in these diseases was unequivocal.

In the mid-1950s, Ernest Witebsky, a follower of Paul Ehrlich and an ardent sceptic of autoimmunity, found that inoculating rabbits with rabbit thyroglobulin, a thyroid protein, led to the production of anti-thyroglobulin autoantibodies in the rabbits. He also saw the infiltration of immune cells into the thyroid glands of the animals. The infiltration of the gland resembled what occurred in a type of thyroid disease in humans first described in 1912 by the Japanese physician Hakaru Hashimoto. In Hashimoto's eponymous disease the thyroid gland was being destroyed by an unknown mechanism. Witebsky tested sera from patients with the disease and found autoantibodies against thyroglobulin. These observations forced Witebsky finally to acknowledge the existence of autoimmunity.[9] In 1956, Ivan Roitt and his colleagues in London titrated the levels of anti-thyroglobulin antibodies and found them to be higher in Hashimoto patients compared to controls.[10] Therefore, in 1957 Hashimoto's thyroiditis became the first human disease to be recognized as an organ-specific autoimmune disorder. However, we now know that these autoantibodies against thyroglobulin are not the main immune mechanism in Hashimoto's disease: immune cells that infiltrate the thyroid gland appear to be responsible for its destruction.

In 1964, an international conference on autoimmunity convened by the New York Academy of Sciences finally reached consensus on the reality of autoimmunity and autoimmune diseases.[iii]

So Ehrlich's *horror autotoxicus* theory was destroyed. Autoimmunity can and does occur. Many chronic diseases are now thought to have an autoimmune aetiology. But how does one go about proving that a certain disorder is an autoimmune disease? Is it enough to demonstrate the presence of autoantibodies against self-antigens? The answer to that question turns out to be a resounding no. Autoantibodies against certain self-antigens, such as those found inside nuclei (anti-nuclear antigens), are found in healthy individuals, especially in the elderly. The best kind of evidence would involve the transmission of the characteristic features of the disease to healthy animals or humans, by transfer of self-reactive antibody or immune cells. Such a transmission could happen naturally, such as by the transfer of autoantibodies from mother to foetus, or experimentally by the injection of the putative autoantigen into a human or other animal. This would be analogous to using Koch's postulates to prove that an infectious disease is caused by a particular microorganism.

Examples of autoimmune diseases that have been transmitted to healthy individuals by autoantibodies – naturally or experimentally

– include idiopathic thrombocytopenic purpura, Graves' disease and myasthenia gravis. In some of these naturally produced diseases the infant shows temporary signs of disease due to transplacental transfer of maternal autoantibody. This is the most compelling kind of evidence as it is natural, and the transmission happens from human to human. Having an animal model of the disease would also be useful, but not all autoimmune diseases can be reproduced in an animal or transferred from humans to animals. But pemphigus vulgaris and bullous pemphigoid, two diseases in which autoantibodies cause blisters by attacking proteins that bind our skin cells together, *can* be transmitted from humans to animals by autoantibody.

Another way to show the damaging effect of an autoantibody is to reproduce the pathological mechanism *in vitro*: in a dish. Pernicious anaemia is an autoimmune disease in which vitamin B12, essential to making DNA, fails to bind to a receptor in the gut called 'intrinsic factor'. Autoantibodies bind to intrinsic factor and block the binding of the vitamin, which then cannot be absorbed. Without enough DNA, red-blood-cell formation is impaired and the patient becomes anaemic. They can also show neurological and psychiatric symptoms such as tingling extremities, balance problems and depression. Autoantibodies from patients with pernicious anaemia can be shown to inhibit the binding of B12 to intrinsic factor *in vitro*.

If the putative autoantigen is known it can be purified and injected into an animal; if the animal then gets the characteristic lesion of the autoimmune disease, this can be considered as *indirect evidence* of the autoimmune nature of the disease. One example is multiple sclerosis, in which the immune system starts attacking the myelin sheaths surrounding nerve cells in the brain; loss of the myelin – demyelination – hampers the ability of the nerve cells to transmit nerve impulses, leading to neurological dysfunction. Patients with multiple sclerosis suffer from a range of neurological symptoms such as weakness, vision problems, loss of balance, urinary incontinence, and so on. Injecting rats with myelin protein will give rise to a demyelinating disease in the animal, indicating the probable role of myelin protein in generating multiple sclerosis in humans. The evidence presented in favour of an autoimmune aetiology for most autoimmune diseases is derived from this kind of experiment.

So how and why does autoimmunity occur? There are several possible aetiologies, but they all involve an important dysregulation of the immune system: for it to occur there must be a *loss of tolerance*: that is, the mechanisms that allow tolerance to self-antigens must somehow be breached.

i   This mechanism is called molecular mimicry. Some pathogens have antigens that are so similar to our own antigens that antibodies and T-cell responses mounted, perfectly reasonably, against these foreign invaders may cross-react with their very similar self-antigens, giving rise to autoimmunity. For example, a simple sore throat caused by a streptococcus can lead to the formation of anti-streptococcal antibodies that attack self-antigens in the heart because of similarity to certain streptococcal antigens. This complication of a streptococcal infection is called rheumatic fever.

ii If a rhesus-negative (Rh–) woman has a foetus which is rhesus positive (Rh+), by virtue of the father being Rh+, then during birth Rh antigen (D) sensitizes the mother resulting in anti-D (anti-Rh+) antibodies developing in the mother. If there is a second pregnancy, and if the foetus is Rh+, these autoantibodies from the mother can cross the placenta and attack the foetus.

iii The chance observation of a serum component from a rheumatoid arthritis patient that agglutinated native gamma globulin coated sheep red blood cells led to the discovery in 1940 of the serum factor, which was later found to be IgM that reacted with the Fc part of IgG. This was named rheumatoid factor.

Chapter 17

INAPPROPRIATE RESPONSES

Toulon, France, 1901. A beautiful yacht, the *Princess Alice II*, its billowing white sails gleaming in the morning sun, sailed out into the Mediterranean Sea. This, however, was no pleasure cruise, but an annual scientific voyage hosted by Albert, Prince of Monaco. On board with the prince, a keen amateur oceanographer, were his guests: a team of scientists accompanied by a menagerie of unfortunate laboratory animals. One of the key figures on this cruise was the Parisian physiologist and physician Charles Richet. Richet and his colleague Paul Portier were interested in studying the poison produced by *Physalia*, the marine creature known as the Portuguese man-of-war. But first they had to find the elusive medusoid, and the first week passed without so much as a glimpse of the long-tentacled creature. Richet decided to use this time on his other great passion – playwriting – and completed a drama, *Circe*, a tragedy about Homer's great hero Ulysses. The play was eventually produced by the prince with Sarah Bernhardt in the leading role.

Charles Richet was born in Paris in 1850. He studied medicine at the University of Paris, gaining his medical doctorate in 1869. A brilliant career in research led to his appointment as professor of physiology at the Faculty of Medicine in 1887. Richet's varied research studies led to the elucidation of the mechanism of thermoregulation in warm-blooded animals, and the demonstration of the toxic effects of heavy-metal salts on various bacterial species. It was his interest in marine poisons that led to the invitation to join the cruise and the collaboration with Paul Portier.

Paul Portier was born in Bar-sur-Seine, France, in 1866. Like Richet, Portier studied medicine, obtaining his medical doctorate in 1897 from the University of Paris. He was an able scientist, gaining a doctorate in science from the same establishment. After completing his studies, Portier became an assistant in the departmental chair's laboratory of physiology, through whom he met and developed a lasting friendship with the Prince of Monaco.

At the beginning of the twentieth century, the consensus was that all immune responses directed against pathogens were always beneficial. This dogma was to be challenged by the two Parisians in July 1901.

As the *Princess Alice II* approached the Cape Verde Islands the Mediterranean finally revealed colonies of *Physalia*. Now Richet and Portier were able to begin their experiments. They soon discovered that the tentacles were the source of the creature's poison: injecting glycerol extracts from tentacles caused profound central-nervous-system depression and death in ducks, guinea pigs and pigeons. After the cruise ended, Richet and Portier decided to continue with their experiments. Back in the laboratory in Paris, they studied the effects of a similar poison, actinotoxin, obtained from the sea anemone *Actinia sulcata*, on dogs. Using 36 dogs they established the poison's 'LD50' (the 'lethal dose' required to kill 50 per cent of the animals).[1]

But another experiment, also carried out in 1901, produced unforeseen results. Eight dogs that had been given a small dose of poison, well below the LD50 and not thought to be large enough to cause any ill effects, began showing adverse reactions immediately after being injected. They started defecating, began to vomit and went into convulsions, before collapsing and dying. Death had come to all the animals within minutes. It was later found that the dogs had been exposed to the toxin in a previous experiment; but instead of becoming immune they had evidently become hypersensitive.[2]

They repeated the experiment with two dogs called Galati and Neptune. The dogs received three small doses of toxin on day one, day eight and day 27. Both animals died after the administration of the third injection on the twenty-seventh day.

The laboratory notebooks state:

10 Feb (1902). 26 days after first injection – the dog was in perfect health, cheerful, active; the coat was shiny. On this day at 2pm it was injected with 0.12 cc toxin per kg. Immediately produced vomiting, defecation, trembling of front legs. The dog fell on the side, lost consciousness, and in one-half hour was dead.[3]

Portier and Richet concluded that an inherent property of the substance had produced sudden, escalating symptoms and death, even though the injected dose was insufficient to kill or even sicken a normal animal. Richet called it 'anaphylaxis' – *a* for 'contrary to' and *phylaxis*, 'protection'. No doubt thinking of Louis Pasteur's famous statement, Portier later commented on their finding: 'We discovered anaphylaxis without looking for it, and almost in spite of ourselves. But it was necessary to have the eyes and mind of a physiologist to understand the interest.'[4] Interestingly, at the time Richet and Portier did not consider anaphylaxis to be an immune phenomenon. After subsequent experiments, however, they appear to suggest that it might in fact involve an aberrant immune reaction.

In 1903, Maurice Arthus, a doctor from Angers, France, working at the Pasteur Institute, observed unusual skin reactions to foreign proteins. Injecting horse serum under the skin of rabbits, he found that after the fourth injection there was a local swelling, which after the fifth injection began oozing pus and after the seventh became gangrenous. The skin reaction was consequently called the 'Arthus reaction'.

A similar type of hypersensitivity had also been observed in some patients following serotherapy vaccination by tetanus or diphtheria antitoxin and was called 'serum sickness'.[5] But were these unusual reactions immune phenomena? The riddle was solved by a paediatrician named Clemens von Pirquet.

*

Pirquet, a Viennese aristocrat, was born in 1874. He studied medicine in Vienna, Königsberg and Graz, obtaining his medical doctorate in 1900. Following graduation, Pirquet trained in paediatrics and soon became one of the leading paediatricians in Vienna. While working at the Kinderklinik in Vienna, Pirquet routinely administered antitoxin serum to children suffering from diphtheria and scarlet fever. Frequently he came across a child who exhibited a strange reaction to the serum: a rash, fever, swollen lymph glands, joint pain. The children were showing signs of serum sickness, and its all-too-frequent appearance was one of the main reasons why serotherapy had fallen out of favour.

Pirquet had a hunch that these reactions were aberrant immune responses. He noticed similarities between the symptoms of serum sickness and those shown by animals injected with certain substances: these animals, instead of generating a protective immune response, reacted violently to the injection, falling ill, sometimes even dying. He found that the signs and symptoms that appeared following these injections or serotherapy coincided with the appearance of antibodies in the blood; he also observed that when children were given a repeat dose of serum the antibody response appeared much faster than after the first dose. Pirquet's genius lay in his ability to see that these antibody responses to antitoxin (and substances injected into animals that gave unusual reactions) were no different to antibody responses mounted against a pathogen, something that was well known at the time: they were always faster and more pronounced when an immunized animal was given a second injection of the same antigen.[6] This is the secondary response that we came across earlier. He wrote:

The conception that the antibodies which should protect against disease are also responsible for the disease sounds at first absurd. This has as its basis the fact that we are accustomed to see the antibodies solely as antitoxic substances. One forgets too easily that the disease represents only a stage in the development of immunity and that the organism often attains the advantage of immunity only by means of disease.[7]

In 1906, Pirquet put forward his theory that the immune response was the cause of hypersensitivity phenomena such as anaphylaxis and serum sickness. This threw the whole field of immunology into disarray. The immune system was regarded as the body's saviour, the antibodies as its weapons. How could a system that had evolved to seek and destroy the pathogen now turn against itself? At the time, antibodies were generally thought to possess one key function: the neutralization of bacterial toxins. And, as the binding and neutralization of toxins had been clearly demonstrated, they were considered wholly protective. But Pirquet challenged the idea that antibodies were always protective. 'In my hypothesis,' he wrote, 'these other antibodies form a new toxic body with the antigen.'[8] As expected, his suggestion that a disease might be, albeit indirectly, due to an antibody was dismissed.

We now know, however, that Pirquet was right in his assertion. Serum sickness and anaphylaxis were caused by aberrant immune responses, and, like responses to some infections, involved antibodies. But serum sickness was mediated by a different type of antibody to that involved in anaphylaxis. In serum sickness antibodies of the immunoglobulin G (IgG) class are formed against the foreign protein – horse-antitoxin antibodies – introduced into the patient, leading to the formation of 'immune complexes' made up of the horse-antitoxin antibody and human-IgG antibodies generated against them inside the patient. These immune complexes then cause pathology by becoming lodged in various tissues, a process that leads to inflammation. In contrast, anaphylaxis – the reactions seen in the studies of Portier and Richet – involves a totally different antibody: immunoglobulin E (IgE), discovered in the 1960s, which binds to the antigen (sometimes called an allergen) and initiates a series of cellular reactions that lead to inflammation, which in some cases can result in death. In both cases the aberrant immune response leads to an inflammatory response that causes damage to the host. Thus, the immune system was shown to be a double-edged sword, and both hypersensitivity and autoimmunity represented the dark shadows of the generally beneficial immune responses.

Can these unwanted and harmful responses be abrogated? In his book *Serum Sickness*, published in 1905, Pirquet describes an interesting experiment that he carried out on himself. Alexandre Besredka, who had taken over Metchnikoff's laboratory at the Pasteur Institute, was attempting to find a way of preventing hypersensitivity in laboratory animals. He had found that by injecting gradually increasing doses of allergen he could eventually make animals tolerate a dose that would have otherwise led to anaphylactic shock. This was the first example of immunotherapy against hypersensitivity in an animal model, which Besredka called an anti-anaphylaxis response.[i]

After reading Besredka's publication, Pirquet decided to try the method on himself, injecting, every six to ten days, gradually increasing doses of diphtheria and scarlet-fever serum into his skin. Carefully assessing the skin reactions, Pirquet observed that the protocol of repeat injections eventually led to a diminution of the inflammatory response.[9] Some kind of desensitization to the antigen in the serum had taken place. Could this be the answer to the problem of hypersensitivity?

Several physicians attempted desensitization, but the method was not without risk, and the first fatal case of anaphylaxis following injection of increasing doses of horse serum was reported in 1910.[10]

*

Leonard Noon, a surgeon working in the laboratory of the Department of Therapeutic Inoculation at St Mary's Hospital, London, carried out the first trial of allergen immunotherapy against pollen hypersensitivity, which causes hay fever. Noon carefully prepared several extracts of pollen and administered subcutaneous injections of increasing doses, finding that it was 'possible in every case to raise the patient's resistance, to a marked degree, within the lapse of a few months'. He tested this 'resistance' by carrying out a provocation test before and after injection. The test involved administering drops of allergen extract of different dilutions onto the patient's eye and then assessing the ensuing inflammatory response by examining the eye and noting the degree of redness. Resistance, indicated by the patient requiring a stronger extract of pollen to cause an inflammatory reaction, was seen following the series of injections carried out over a few months. Publishing his findings in *The Lancet* in 1911, Noon attributed his success to finding the right dosage of pollen extract.[11]

Tragically, Noon, the pioneer of allergen immunotherapy, succumbed to tuberculosis and died aged just 33. But his work was continued by John Freeman, who tested Noon's 'pollen vaccine' on a larger cohort of patients, obtaining similar results.[12] Noon and Freeman's method of eliciting this lack of responsiveness to pollen – or 'hyporesponsiveness' – was enthusiastically adopted by many physicians throughout the world. The first controlled clinical trial was carried out in 1955 in England by William Frankland, who demonstrated that immunotherapy against hay fever was more effective when high doses of pollen extract were used.[13]

In the 1970s allergenic proteins were modified by formaldehyde or glutaraldehyde to produce allergoids, which, like toxoids, were safer and less likely to give anaphylactic reactions.[14] Allergoids, like toxoids, were immunogenic but not allergenic, in the sense that they did not provoke adverse reactions. Importantly, they conferred resistance against hypersensitivity to the original allergen.

Immunotherapy has also been successful in conferring protection against insect-venom hypersensitivity, thereby preventing lethal anaphylactic reactions that can follow bee or wasp stings in individuals allergic to the venom. In 1953, Mary Loveless and John Cann found that beekeepers, who by the nature of their work were being immunized by bee antigens daily, did not develop hypersensitivity to the insects' venom. Interestingly, they were able to transfer this protection through beekeeper serum to sensitive individuals, indicating that a serum component was responsible for conferring protection.[15] This component was found to be an IgG antibody that, by binding to bee allergen, effectively blocked the allergen binding to pathogenic IgE antibodies that mediated the allergic reaction.

Pirquet, whose discovery had revolutionized immunology, refused an offer to join the Pasteur Institute, opting instead to work at the Rockefeller Institute in New York. After a year there he returned to Austria and took up the post of professor of paediatrics at the University of Vienna. He continued his research, discovering a new skin test for tuberculosis. But the storm clouds were gathering, and Clemens von Pirquet's story did not have a happy ending. Despite being one of Vienna's most famous paediatricians and possessing a brilliant and insightful research mind that had made a truly original discovery of the pathological nature of certain immune responses, Pirquet suffered from a disastrous personal life. His wife, Maria, to whom he was devoted, was unstable and had developed an addiction to barbiturates. In 1929, without warning, Pirquet and Maria committed suicide by cyanide poisoning. He was just 54 years old.

Soon it became clear that allergic reactions to innocuous substances were common among the general population. Sometimes these hypersensitivity reactions resulted in anaphylactic shock, which was often fatal. What was required was a simple but effective test that could identify those individuals who were allergic to a given substance. Two physician–scientists, this time from the German school, carried out an interesting experiment that led to the development of such a test.

*

Otto Carl Willy Prausnitz was born in Hamburg in 1876. After attending the universities of Leipzig, Kiel and Breslau, Prausnitz gained his medical doctorate from Breslau in 1903. A student of Richard Pfeiffer, he initially trained in bacteriology, before switching fields to allergic diseases. In 1905, Prausnitz took up a research post in London and began studying the role of grass pollen in hay fever. Heinz Küstner, born in 1897, and an obstetrician and gynaecologist by training, was Prausnitz's assistant, and together they carried out their now famous experiment – one that involved using each other as guinea pigs.

Küstner could not eat certain types of fish: he developed severe reactions to them. In their experiment, carried out in 1921, Küstner injected a rather apprehensive Prausnitz with a sample of his serum. Then, 24 hours after giving him the injection into a fold of skin on the abdomen, Küstner applied fish extract to Prausnitz's skin. Immediately the skin became inflamed at the injection site. This suggested that antibodies against the fish antigens, present in Küstner's serum, had, upon being transferred to Prausnitz, reacted with the fish antigens applied to the skin. This had then led to an inflammatory reaction. This observation formed the basis of the 'Prausnitz–Küstner' or 'PK test': a procedure that could be used to ascertain whether a patient had pathological antibodies against a suspected allergen.[16] A test is deemed positive if a skin reaction develops in a healthy person when challenged by an allergen *after* being given an injection of serum from a patient who is suspected of being hypersensitive to the same allergen.

The allergen-specific antibodies transferred from Küstner to Prausnitz were later found to be of the IgE class.[17] The inflammatory reaction, described as a 'wheal and flare', confirmed that the patient was allergic to the substance tested. The PK test was later replaced by the safer skin test, which did not involve injecting serum from patients into healthy volunteers.

But what caused the wheal and the flare? It was found to be down to leaky blood vessels that increased in diameter. Fluid leaving the dilated and leaky capillary network caused the swelling. But what caused this to happen? The key substance that mediated this reaction, histamine, was characterized by an Englishman who went on to win the Nobel Prize for work on nerve transmission.

*

Henry Dale, born in 1875, obtained his medical degree from Cambridge in 1909. Six years earlier, he had spent four months working with Paul Ehrlich in Frankfurt. Dale's interests lay in pharmacology, and he obtained a research post at the Wellcome Physiological Research Laboratories in 1904.

The chemist George Barger had discovered naturally occurring histamine, and, in 1913, Dale found that histamine caused the contraction of previously sensitized uterine strips, a finding that he described as 'one of my most fortunate accidents'.[18] Importantly, the strips showed a similar contraction when exposed to the specific antigen, and it seemed likely that histamine was acting as a key mediator in immune responses. Injection of histamine into the bloodstream resulted in anaphylactic shock in animals, suggesting a role in allergic reactions. Dale's work paved the way to the elucidation of the key role of histamine as a principal mediator of anaphylaxis.

In 1963, Philip Gell and Robin Coombs classified all aberrant immune responses or hypersensitivities into four types.[19] This classification is still used today. Later, a fifth type of hypersensitivity, stimulatory hypersensitivity, was also described.

Types I, II, III and V all involve inappropriate antibody responses that lead to an adverse reaction in the host. In contrast, type IV hypersensitivity involves cell-mediated responses that inadvertently cause damage to the host. In type I hypersensitivity, described by Portier and Richet, Henry Dale, and Küstner and Prausnitz, the culprit is an IgE antibody, discovered in the late 1960s by Teruko and Kimishige Ishizaka.[20] So what is happening here?

When a genetically predisposed individual or animal first comes across an allergen, by ingestion, inhalation or injection, they become sensitized. This is what happened to poor Neptune and Galati: the dogs became sensitized to the plant toxin when they were given the first injection. When the sensitized individual or animal is re-exposed to the allergen, a full secondary immune response ensues, but instead of being protective, this turns out to be an aberrant reaction mediated by IgE. This, as we saw earlier, can in some cases cause a deadly anaphylactic shock.[ii]

Type II hypersensitivity also involves antibodies, usually of the IgG class, that cause unwanted effects in the host. A good example is erythroblastosis fetalis or haemolytic disease of the newborn, which we came across earlier. The anti-D antibodies can enter the foetal circulation through the placenta and bind to the D antigen on the foetal red blood cells. Complement then binds to these antibody-coated red cells, causing their lysis and resulting in haemolytic disease. The same thing can happen if incompatible blood is transfused. In autoimmune haemolytic anaemia, antibodies against red cell antigens are formed, and in Goodpasture's syndrome kidney and lung antigens give rise to pathological autoantibodies. The result is tissue damage. As you can see, some of these hypersensitivity diseases are autoimmune diseases!

The third type of antibody-mediated hypersensitivity (type III hypersensitivity) involves antigen–antibody complexes. This is what is observed in serum sickness and in the Arthus reaction. Here the antigen – which could very well be a horse-antitoxin antibody – binds with an antibody that is made in the body against the foreign protein, and the immune complex that is formed leads to inflammation.[iii]

Type IV hypersensitivity involves immune cells. Examples include Koch's tuberculin reaction and contact dermatitis, a skin inflammation caused by immune responses against certain substances such as poison ivy and nickel. The mechanism is like the other responses in that there is a sensitization stage followed by a second encounter in which the tissue reaction is observed. The mediators released in the effector phase cause inflammatory cells to migrate into the tissue, and this causes a delay in the appearance of signs and symptoms; hence type IV hypersensitivity is called a 'delayed-type' reaction. You will recall that Portier and Richet's dogs developed symptoms rapidly, whereas Koch's guinea pigs took 48 hours or so to show the skin reaction.[iv]

In 1835, the Irish physician Robert James Graves described a disease in which patients presented with a swelling of the neck (a goitre), a fast heart rate and widened, bulging eyes.[21] This disease, which came to be called Graves' disease, causes the thyroid gland to become overactive. The disease mechanism involves an autoantibody binding to thyroid-stimulating hormone receptors on the thyroid gland, thus mimicking the action of the thyroid-stimulating hormone released by the pituitary gland. This stimulation results in the overproduction of thyroid hormones, which causes the symptoms characteristic of an overactive thyroid gland. These, in contrast to previously mentioned antibodies, are stimulatory in nature. Hence this type of hypersensitivity is called 'stimulatory' or 'type V hypersensitivity'.

i   Exactly how hyporesponsiveness occurs has not been fully elucidated. One theory involves the formation of blocking antibodies generated by signals provided by regulatory T cells that have themselves been generated by immunotherapy. Hence IgG or IgA antibodies are formed which, by binding to allergen, prevents their binding to allergen specific IgE antibodies. Regulatory T cells have also been shown to suppress eosinophils, mast cells and basophils, cells involved in hypersensitivity reactions.

ii The mechanism of type I hypersensitivity begins with the first exposure to an allergen, such as pollen protein. Allergen is captured by antigen-presenting cells like dendritic cells and B cells, processed and presented, bound to the major histocompatibility class II antigen to a naive T helper cell (T helper 0 cell). The Th0 cell differentiates into a T helper 2 cell: the cytokine IL-4 appears to be required in the cellular environment for the naive T helper cell to become a T helper 2 cell. The helper type-2 cell, by secreting more IL-4 and IL-13, promotes class switching in B cells, enabling the production of antigen-specific IgE antibodies. These antibodies bind to high-affinity receptors on mast cells and basophils. The animal or human is then deemed to have been sensitized. Re-exposure to the allergen results in the allergen binding to and cross-linking the bound IgE antibodies on the mast cells and basophils. This cross-linking sends signals to the cell interior, causing cytoplasmic organelles called granules to release their contents, a process called 'degranulation'. Degranulation thus causes primary mediators to be discharged into the tissues, where they cause a myriad of inflammatory sequelae such as increased vascular permeability, extravasation of cells, granulocyte chemotaxis and smooth muscle contraction. In the case of someone suffering from hay fever, this would normally manifest itself as a local reaction, the inflammation giving rise to a runny nose, tearing, coughing, and so on. Sometimes a generalized type I hypersensitivity reaction can happen following ingestion of nuts or seafood, insect bites (venom) and drug injections in sensitized individuals. On re-exposure these individuals will become hypotensive, short of breath, wheezy, and have swollen lips/eyelids and an urticarial rash. Death in such cases occurs due to the systemic release of vasoactive mediators, leading to blood-vessel permeability and dilatation throughout the body, resulting in sudden loss of blood pressure, massive oedema – which can cause swelling of the voice box and impair breathing – and severe bronchiole constriction which can impair respiration.

iii In post-streptococcal glomerulonephritis, immune complexes are formed following a streptococcal infection, like strep throat. The antibodies bind to antigens from certain streptococci, and the complexes become lodged in the glomerular basement membrane of the kidney, below the podocyte foot processes. These antigen–antibody complexes can be seen as 'lumpy bumpy' structures on light microscopy and sub-epithelial humps on electron microscopy. The immune complexes cause complement activation, which generates an inflammatory response, which in turn leads to the destruction of the basement membrane.

iv In the sensitization phase, allergens are captured by antigen-presenting cells (APCs) such as the Langerhans cells (dendritic epidermal cells), and these are presented to T cells. On re-exposure to the allergen the sensitized cells expand and release cytokines which cause inflammation. Some of these cytokines also recruit inflammatory cells to the site, a process that can take up to 72 hours. This causes the delay in symptom appearance. The sensitization process can be brief (six to ten days for strong sensitizers such as poison ivy) or prolonged (years for weak sensitizers such as sunscreens, fragrances and glucocorticoids).

Contact dermatitis is an example of type IV hypersensitivity. The chemicals that can provoke allergic contact dermatitis are small molecules (<500 daltons), and approximately 3,000 such chemicals have been recognized as being capable of causing allergic contact dermatitis. The small chemical molecules responsible for allergic contact dermatitis must be bound to carrier proteins in order to provoke an immune response.

Summary

THE END OF THE BEGINNING

This concludes the first period in the history of the immune system, from antiquity to the early twentieth century. So, what did we know about the immune system by the end of this first period?

Early observations established the principle of immunity and that individuals who had contracted an infection were able to resist it if they came across it a second time. The cellular basis of this protective secondary response was of course unknown, but the principle was exploited in many early interventions such as variolation and Jennerian vaccination. The emergence of the germ theory of disease and the identification of causative agents of several infectious diseases – the work of Pasteur and Koch – led to the further development of vaccines, with the first controlled study being carried out at Pouilly-le-Fort.

But who were the foot soldiers of this immune army? Antibodies, in the form of antitoxins, were discovered by Behring and Kitasato, and this breakthrough led to a novel method of treatment, serotherapy. Ehrlich then put forward his side-chain theory, which attempted to explain how antibodies were formed in response to a toxin; this was the first selective theory of antibody formation.

Metchnikoff, by describing phagocytosis, first proposed a role for cells in immunity. Koch's discovery of the delayed-type hypersensitivity reaction suggested that cellular factors were involved in immune responses to certain infections such as tuberculosis. Soon two camps emerged: the humoral camp, propounding the importance of serum antibody; and the cellular camp, insisting on the vital role of phagocytic cells in immunity.

Metchnikoff also demonstrated that the inflammatory response, thought to be solely pathological at the time, was a key innate defence mechanism.

Pfeiffer's description of bacteriolysis – the bursting of bacterial cells – showed that antibodies were not only antitoxin moieties but also able to cause the lysis and destruction of microbes. Serum factors, such as complement and binding antibody, were then shown, by Bordet and Wright, to be important bridges connecting the humoral and cellular responses. Natural antibodies reactive against red-blood-cell antigens, also described in this period, led to the classification of the human ABO blood system.

It soon became apparent that immunity was indeed a double-edged sword, and that harmful immune responses were, in some cases, generated against innocuous substances. Thus, allergies and other hypersensitivities were described, and, contrary to Ehrlich's *horror autotoxicus* theory, it was shown that autoimmunity unfortunately did happen in certain cases. Finally, Coley's work showed that immune responses could also be unleashed against certain cancers.

It is also interesting to acknowledge what we *didn't* know at the end of that period. We didn't know the nature of antibody molecules, how they were formed or the mechanism that generated their tremendous diversity against a seemingly endless repertoire of antigens. We also didn't know anything about the cells involved in the specific immune responses: the mysterious lymphocytes. We didn't know about cytokines, which mediated communication between immune cells. We didn't know about the existence of immunodeficiency disorders and their varied mechanisms.

We have always known that some individuals are more susceptible to certain diseases than others, but the mechanism underlying this difference was unknown. Was it down to genetics? We did not know. We knew that the immune system generally did not attack self-antigen, but that autoimmunity happened. But how the immune system differentiated between self and non-self remained a mystery.

PART TWO

THE GOLDEN AGE

Chapter 18

THE THINKING RADISH

Oxford, 1941. Sitting in the garden of his home, Peter Medawar and his family heard an approaching plane. Soon they saw the dark shape of a twin-engine bomber flying perilously low towards the house. Running into a shelter, they heard it crash into a nearby house only 200 yards away. As Medawar emerged from the shelter and began running towards the plane it exploded. Medawar wasn't close enough to sustain any injuries, but the airman was severely burned: he was rushed by ambulance to the Radcliffe Infirmary.

Europe itself was on fire: the Second World War was at its zenith. Germany had once again conquered its nemesis and Paris was an occupied city. But in England, Medawar, a young zoologist at the time, became preoccupied with a central question in immunology. How did the immune system differentiate between 'self' and 'non-self'? Even though he was aware of the reality of autoimmunity, he inferred the existence of a mechanism or mechanisms that generally prevented the immune system from attacking self-antigen. In other words, the immune system had somehow developed or acquired tolerance to self-antigen. But how?

Peter Medawar was born in 1915 in Rio de Janeiro, Brazil, to a Lebanese father and English mother. After the First World War, his family moved to England. After attending Marlborough College, Medawar entered Oxford University, where he obtained a first-class honours degree in zoology in 1935. He stayed on at Oxford, working as a senior demonstrator in zoology before venturing into medical research under Howard Florey at the School of Pathology.

Following the crash, Medawar decided to visit the injured pilot at the infirmary. The man was in bad shape, with extensive third-degree burns. During his visit, Medawar met Dr John Barnes, the physician who was treating the patient. In his autobiography Medawar recalls what the clinician, upon discovering that Medawar was a research biologist, said to him. 'He told me to lay aside your intellectual pastimes and take a serious interest in real life.'[1] The pilot needed a skin graft, a notoriously unpredictable treatment, and the clinician was in effect asking Medawar to come up with a solution that might help improve the survival of the grafts, and therefore of the airman.

At the time, it was known that grafts between different individuals of the same species (allografts or homografts) would be rejected – most of the time – unless they were between identical twins or from inbred strains of laboratory animals (isografts).[i] Isografts *took*: in other words, they were tolerated. Medawar wrote to the War Wounds Committee of the Medical Research Council and managed to obtain a grant to study skin grafting. He was dispatched to the Burns Unit at the Glasgow Royal infirmary, where he began working, as he states, 'under the direction of Dr Leonard Colebrook, one of the very distinguished pupils of Sir Almroth ("stimulate the phagocytes!") Wright'.[2] At the Burns Unit, Medawar teamed up with the Scottish surgeon Tom Gibson, a collaboration that was to lead to the establishment of the scientific basis of graft rejection.

Their first study involved a single patient, 'Mrs McK', an epileptic who had severely burned her chest after falling against a gas fire. Medawar and Gibson applied pinch grafts on the burns using skin taken from the patient and from her brother. At regular intervals following the grafting they took small button specimens of the grafts by lifting a small piece of the skin and making a horizontal incision. Examining them under the microscope, they found that the grafts were infiltrated with cells which interestingly included lymphocytes. The key finding that Medawar made from this study involved the observation that second grafts from Mrs McK's brother were rejected much faster. This indicated an immunological secondary response, just like one mounted against a pathogen.

In their paper, 'The fate of skin homografts in man', the authors write:

> Theories about the mechanism that underlies resistance to homoplastic grafting are of two general types. The first is that the reaction is primarily local and cellular, and that evidence for the failure of homografts is to be found within the grafts themselves or in their immediate neighbourhood. This view is based mainly on histological considerations... The second, long familiar to students of tumour transplantation, is that resistance to homoplastic grafting is systemic and primarily humoral in nature, and that in one form or another it follows the general pattern of an antigen–antibody reaction. This view is in accord with general serological theory and is based on the idea that the proteins and possibly other ingredients of the grafted tissues are sufficiently unlike those of the recipient to act as iso-antigens. The observations recorded in this paper favour this second hypothesis.[3]

Clearly Medawar believed at the time that the immune response that caused graft rejection involved antibodies. Humoral theories still held sway, and antibodies were considered to be the predominant factors in immunity. He looked for these anti-graft antibodies, but found none.

Medawar returned to Oxford and began what he calls in his autobiography the 'hardest stint of work I had ever undertaken in my life'.[4] He was trying to unravel the underlying basis of graft rejection. In one experiment, he carried out skin grafts between 24 rabbits, a total of 600 grafts in total. By looking at the pattern of rejection in these animals Medawar showed that transplant rejection was under genetic control.[5]

At the International Congress of Genetics in Stockholm, Medawar met Hugh Donald, head of the Agricultural Council's animal-breeding research unit in Edinburgh. Donald's research involved cattle twins. Their amiable conversation drifted into a discussion about how one could distinguish between fraternal and identical-twin cattle: fraternal 'two-egg' twins were genetically dissimilar animals, whereas identical twins were produced from a single fertilized egg, which subsequently divided to give rise to 'one-egg' twins.

Medawar suggested a way of distinguishing between fraternal and identical twins. In Medawar's own words, he said, clearly rather confidently:

> In principle the solution is extremely easy: just exchange skin grafts between the twins and see how long they last. If they last indefinitely you can be sure these are identical twins, but if they are thrown off after a week or two you can classify them with equal certainty as fraternal twins.[6]

He offered to demonstrate his skin-grafting technique to Donald's veterinary staff. A few months later Donald wrote to Medawar reminding him of his promise, and Medawar accompanied him to an experimental farm near Birmingham to carry out skin grafts on the twin cattle. But, as Medawar wrote, 'The results were not at all what we had expected.'[7]

The skin grafts, they found, were *accepted* by the two-egg, fraternal twin animals, and this puzzled Medawar. These were genetically different animals, so why were the skin grafts not rejected? There appeared to be some form of *tolerance* to the graft, and this tolerance was *specific* to the twin: sibling or parental grafts were rejected, and only grafts from the other twin were accepted. What was going on?

Then Medawar recalled a study done in 1945 by the American geneticist Ray Owen. In Owen's experiments two-egg twin cattle were found to be *chimeras* for red cells. Because of the connection of their circulatory systems within the womb, with the subsequent mixing of blood, each twin contained red blood cells from the other.[8]

Owen's study gave birth to a hypothesis, one that might explain why grafts between fraternal 'two-egg' twin animals were tolerated by each of them. Medawar believed that during foetal and newborn life, self-markers on the surface of cells were recognized as such by the immune system, leading to a tolerance: an absence of an immune response against these antigens. Therefore, if a foetus or newborn animal were exposed to a foreign antigen during this 'window period', it would be considered 'self'. Due to the connection of their circulatory system, each of Ray Owen's fraternal cattle twins had become tolerant to the other's antigens. This was Peter Medawar's great discovery: acquired tolerance.

Acquired tolerance could explain another observation made far back in the 1930s: a mouse foetus infected with lymphocytic choriomeningitis (LCM) virus from the mother through the placenta was unable to mount immune responses against the LCM virus when later challenged after birth. Therefore, these mice had developed an acquired tolerance to LCM-virus antigens. It all made perfect sense to Medawar. He just had to prove it.

Now based at University College London, Medawar and his colleagues, Rupert Billingham and Leslie Brent, designed an experiment to test this hypothesis. It involved the use of two genetically different strains of inbred mice: a brown strain (called CBA) and a white strain (called A). Normally these animals would reject skin grafts from each other. In their experiment, they injected cells – spleen, testes, kidney – from a strain-A mouse into a brown CBA-mouse foetus. Once the CBA mouse was born, skin from a strain-A mouse was grafted onto the CBA mouse. To their delight, Medawar and colleagues found that the graft *took*: it survived; it was not rejected. This experiment duplicated the immunological phenomenon that naturally occurred in Ray Owen's twin cattle. The CBA mouse had acquired tolerance to strain-A antigens by being exposed to strain-A cells while in the womb. The strain-A cells injected into the strain-CBA foetus were considered 'self' by the immune system of the CBA foetus, and when it was born it could happily accept a skin graft from a strain-A mouse.

Subsequent experiments found that, like foetuses, newborn mice could, by the intravenous transfer of donor lymphocytes, be rendered tolerant to skin grafts from genetically different mice. However, this could only be achieved if the lymphocyte transfer was done within a window period of a few days after birth, after which such grafts were rejected.

The results of this work were published in 1953, in the internationally renowned journal *Nature*.[9] Medawar had discovered that when the immune system distinguished 'self' from 'non-self', it all seemed to happen during foetal and early neonatal life. In 1960, Peter Medawar was jointly awarded the Nobel Prize for Physiology or Medicine for the discovery of 'acquired tolerance'. The other laureate was Frank Macfarlane Burnet, whom we shall meet later.

In 1969, while delivering a lecture to the British Association for the Advancement of Science, Medawar suffered the first of several strokes, which left him partially paralysed on the left side of his body, but upon recovery he went back to work. He even attempted, for a while, to continue doing benchwork – which he describes as a 'great pleasure' – using his right hand with an assistant playing the part of his left hand.[10]

It is important to understand what we mean by tolerance. Tolerance refers to the *specific* immunological non-reactivity to an antigen resulting from a previous exposure to the same antigen. The most important form of tolerance involves non-reactivity to self-antigens, which happens during foetal and early newborn (neonatal) life. The exact mechanism of induction and maintenance of tolerance is not fully understood.[ii]

Medawar's theory suggested a selection process occurring during foetal and early neonatal life. During this critical period, immune cells are exposed to self-antigens, and cells capable of reacting with self-antigen are somehow deleted or made inactive. This mechanism would safeguard against rampant autoimmunity. Conversely, if the immune system did not 'see' an antigen, during this period it would not be tolerant to it even if it was a self-antigen. What would happen if the immune system came across this missed self-antigen later in life? As far as the immune system is concerned, this would be a foreign (non-self) antigen – like one belonging to a pathogen or genetically dissimilar graft – which needed to be eliminated. Now, it so happens that some of our self-antigens are hidden from the scrutiny of the immune system during embryonic and neonatal life. These may be antigens that only appear later in life, after the neonatal period, or are confined to specialized organs like the testes, brain or eye. These latter antigens are said to be sequestered, shielded from the selection process that enables the development of self-tolerance. So what would happen if some of these sequestered antigens – say, from an eye – were released into the circulation by a traumatic injury? They would be recognized as foreign and attacked, and, in the case of the eye, the immune response would target antigens in both eyes even though the trauma only involved one.

Medawar, Billingham and Brent came across another strange phenomenon. They found that an animal that had been made tolerant by injection of foreign cells when a foetus, and then given a skin graft from the donor, was sometimes attacked *by the graft*, causing runting or wasting disease in the recipient.[11] But this only happened if the graft – which, as expected, was not rejected – contained lymphocytes. The lymphocytes in the donor graft were attacking the host. This unusual response was called a 'graft-versus-host response', and it appeared to be very similar to the delayed-type hypersensitivity response originally described by

Robert Koch. Koch had observed an inflammatory reaction when cells from one animal were injected into the skin of another; a similar reaction was found when the experiment was carried out in human volunteers. Here was yet another mystery that had to be solved!

In 1948, a young zoology graduate, Avrion Mitchison, joined Medawar's team at Oxford to carry out research towards a doctorate. Mitchison was working on mouse tumours, lymphosarcoma, that – like skin grafts – were also rejected when introduced into a genetically different strain. To his surprise, Mitchison observed an accelerated rejection of a tumour given to a mouse that had already rejected a tumour from the same strain. This heightened response to a second challenge was a hallmark of immunity: the mouse was clearly mounting an immune response against the tumour. Mitchison also noted that mice transplanted with tumours had enlarged lymph glands, also referred to as lymph nodes.[12]

Mitchison wanted to find out which immune component was responsible for the rejection. Was it the cellular compartment, or antibodies? To find out, he designed an elegant experiment. First, he injected lymph-node cells taken from a mouse that had recently rejected a tumour into a normal mouse. Then he challenged this second mouse with a tumour: the tumour was rejected much faster. This suggested that sensitized cells from the first mouse were responsible for the accelerated rejection. However, if he challenged the second mouse with a tumour after injecting *serum* from the mouse that had rejected the tumour, the tumour was not rejected any faster.[13] This proved that it was lymphocytes, the cells found in the lymph nodes, not antibodies, that mediated tumour rejection.

Medawar was taken aback when he saw these results. He, like everyone else at the time, believed that rejection was antibody-mediated even though the anti-graft antibodies remained elusive. Immediately Medawar, Billingham and Brent repeated Mitchison's experiment using skin grafts and obtained the same results. Acute graft rejection, it appeared, was mediated by lymphocytes.

Later experiments carried out in 1954 confirmed these findings. Grafts placed inside a chamber impermeable to cells, but not fluids, and implanted into animals evaded rejection even if the host was previously sensitized.[14] The delayed-type hypersensitivity reaction was also confirmed to be transferable by cells.

However, graft rejection in genetically dissimilar ('allogeneic') animals from the same species was not absolute, and a technique called 'tissue matching', if carried out, would allow such grafts between animals from the same species to be accepted. We shall look at this in more depth in the next chapter.

i   Grafts can be classified into four types. 'Xenografts' are grafts carried out between members of different species. Typically, these will get rejected very quickly as the antigens found on these grafts will be very 'foreign' to the host and a vigorous immune response will be mounted against them. 'Allografts' (referred to as 'homografts' in Medawar's time), the most common type of graft used in clinical practice, are transplanted between genetically different members of the same species. The graft can be from a living donor or from a cadaver. These will also be subject to an immune-mediated rejection, the ease of rejection depending on the degree of difference between certain histocompatibility antigens of the donor and recipient. Tissue matching is therefore undertaken to minimize such rejection. 'Autografts' involve the transplantation of tissues (usually skin) from one part of the body to another in the same individual. Such grafts are not rejected because no foreign antigens are involved. Similarly, 'isografts', which are grafts between genetically identical individuals (e.g. monozygotic twins), will also undergo no rejection.

Pure-bred mice can freely exchange grafts without rejection. Medawar's interbreeding experiments showed that the genes coding for the antigens involved in rejection segregated in a Mendelian fashion. For example, let's take two strains of mice, A and B, that are homozygous for two transplantation antigens, a and b, respectively: they would thus have the genotypes aa and bb. Crossing A and B mice gives first-generation (f1) progeny mice with the genotype ab and, as these mice can accept grafts from either parent, the genes a and b would be codominant.

If we then crossed two of these f1 mice, we would get three out of four f2 mice being able to exchange grafts from either parent. If, however, there were n genes involved in transplant rejection, the proportion of the f2 generation accepting grafts from either parent would be $\frac{3}{4}n$.

Therefore, by such breeding experiments we can estimate the number of genes involved in transplantation rejection. In the mouse 20 such genes were found. One of these genes, a gene complex, was found to contain the most important genes responsible for graft rejection. This complex coded for antigens that provoked vigorous immune responses, so these antigens determined whether a graft from one individual to another was rejected or accepted. This gene complex is called the major histocompatibility complex: the H2 locus in the mouse and the HLA locus in man.

ii   There are two distinct types of tolerance: central and peripheral. Central tolerance involves the deletion of any self-reactive lymphocytes. In mammals this happens in the thymus and bone marrow. Central tolerance is not a foolproof mechanism: autoreactive cells often fail to undergo deletion and can potentially cause autoimmune disease. To circumvent this problem there are additional checkpoints in place that attempt to ensure the maintenance of self-tolerance: these mechanisms constitute peripheral tolerance. Peripheral tolerance mechanisms operate after lymphocytes mature and enter the periphery. An important mechanism of peripheral tolerance involves the generation of regulatory T cells (Treg cells) that suppress autoreactive immune responses. These regulatory T cells – formerly known as suppressor T cells – attempt to maintain immunological tolerance by switching off T-cell-mediated immune responses suppressing potentially autoreactive T cells that may have escaped the process of central tolerance in the thymus. Tregs that express forkhead box protein P3 (FOXP3) can transfer immune tolerance, especially self-tolerance, and as such might be useful in the prevention of graft rejection. In animal studies, the induction or administration of FOXP3-positive T cells has led to marked reductions in the severity of several autoimmune diseases such as diabetes, multiple sclerosis, inflammatory bowel disease and thyroiditis. Mutations of the FOXP3 gene, by preventing the development of FOXP3+ regulatory T cells, cause the fatal autoimmune disease IPEX.

Another mechanism that serves to limit autoimmune responses involves the generation of hyporesponsiveness (or 'anergy') in lymphocytes which encounter self-antigen. This happens because there is an absence of co-stimulatory signalling when self-antigens are presented to T cells. In contrast, co-stimulatory signals are present when foreign antigens are presented to T cells.

Chapter 19

THE GIFT

New York, 1958. Dr Donnall Thomas watched as they brought the little girl into the ward. The three-year-old had developed a haematoma, a swelling of blood, after a fall at home. Tears welling in his eyes, Thomas examined the child: she was in a coma. In her short life the girl had suffered greatly. Diagnosed with acute leukaemia, she had been given several bone-marrow transplants from her twin sister. Prior to this procedure, which had never been done before, she had had to endure so much: she had been given powerful immunosuppressive drugs and numerous blood transfusions, and before the grafting her whole body had been subjected to high doses of radiation in order to destroy the cancerous cells that caused the leukaemia. It was hoped that grafting her with bone-marrow cells from her twin sister would repopulate her blood system and save her from the blood cancer.

Thomas recalled the first time he had seen the girl. She had been admitted to the Harriet Lane Home hospital some five months previously after presenting with swollen eyes and lips, vomiting, night sweats and fever. She was treated with immunosuppressive drugs and steroids, but the leukaemia had not responded. It was then decided that radiation treatment followed by a bone-marrow transplant would be her only chance of survival. Marrow aspirated from her twin sister's breast, thigh and hip bones was frozen and later thawed and infused. The girl had done well for around three weeks, but had then started to run a low-grade fever. As the leukaemia had not completely disappeared from her blood she had been given another bone-marrow graft after a further dose of whole-body irradiation. This was repeated: the girl had to endure three bone-marrow transplants in total. But 48 days after the last transfusion blast cells were again seen in her blood: the leukaemia had returned. She was finally discharged five days later.

Now she was back. Donnall Thomas knew that there was little he could do for her. Four days after being admitted the little girl died.

A haematologist by training, Thomas had been fascinated by Medawar's work on acquired tolerance and graft-versus-host disease. He had entered the field in 1949 while doing his residency at Peter Bent Brigham Hospital in Boston.

It was known at the time that the bone marrow produced all the cells found in blood, which included cells of the immune system. Therefore, if the bone marrow of an animal was subjected to high-dose radiation, all these cells, including any leukaemic cells, if present, would be destroyed, and unless the cells were replaced the animal would die.

But early studies had shown that rescue from lethal irradiation could be obtained by infusing irradiated mice with bone marrow taken from a mouse of the same strain: cells in the infused marrow had repopulated the blood.[1] Interestingly, studies carried out in 1955 showed that some lethally irradiated mice could also be saved by bone-marrow infusions from genetically dissimilar mice: the recipients were able to accept skin grafts from the donor.[2]

These studies prompted Thomas, in 1957, to carry out his first set of human 'experiments'. Perhaps grafts from unrelated donors might take? Six patients suffering from leukaemia were given irradiation and chemotherapy followed by infusions of marrow from an unrelated donor. But none of the patients survived beyond 100 days.[3] What had gone wrong? Perhaps the grafts had been rejected?

Then, in 1959, Thomas carried out bone-marrow grafts between identical twins. Two such patients are described in Donnall Thomas's original publication: one, as we saw earlier, died, but the second patient was still alive at the time he was writing. Both patients, however, had a recurrence of their leukaemia. 'Evidently something more than radiation is needed to eradicate leukaemia,' Donnall remarked.[4]

Further studies found that a combination of chemotherapy and X-ray irradiation was more effective at destroying all leukaemic cells before transplantation. These studies also suggested that bone-marrow grafting was more suitable for a patient in remission: with a smaller number of cancer cells it was easier to eradicate all the tumour cells by chemotherapy and irradiation.[5]

Interestingly, he also wrote that 'the twins described in this report were assumed to be identical based on their appearance and identity of blood types. Their histocompatibilities were not verified by skin transplantation.'[6] Thus it appeared that the key to successful allo-geneic bone-marrow transplantation lay in careful tissue matching. If a skin graft from the donor had been carried out and it had not been rejected, the donor and recipient were deemed histocompatible even though they were genetically non-identical.

Donnall did his key experimental work on dogs. In the late 1950s he found that dogs subjected to massive doses of irradiation would recover if the animal was given an infusion of its own bone marrow that had been harvested prior to irradiation. These studies also showed that marrow could be frozen for prolonged periods of time and still be used successfully to reconstitute the dog's entire blood-cell or haematopoietic system. Commenting on these early findings, Thomas concluded, however, that 'in the dog, as in our human patients, marrow from an allogeneic donor almost always resulted either in failure of engraftment or in successful engraftment followed by lethal graft versus host disease'.[7] Therefore, as expected, marrow transplants from a genetically dissimilar donor did not usually *take* and could even attack the host: the graft-versus-host reaction. The only way they would be tolerated was by subjecting the donors to lifelong immunosuppressive drugs.

But he did notice that an occasional dog, usually a littermate, *was* able to tolerate an allogeneic graft. More work followed, and Donnall found that the key to successful graft survival lay in careful tissue matching between donor and recipient. Once the technique for tissue matching in the dogs was developed, Thomas was able successfully to carry out bone-marrow grafting between outbred dogs. Therefore, grafts need not always be from genetically identical animals, as long as tissue matching was carried out.[i]

The two central problems that were encountered in bone-marrow transplantation were an immune-mediated rejection of the graft and a graft-versus-host reaction. Irradiation with 'supra-lethal' doses of radiation and new chemotherapeutic drugs such as methotrexate and cyclosporine somewhat helped to overcome these problems. However, these treatments still rendered the patient immunosup-pressed and at risk of serious infection.

In the treatment of some cancers a patient's own marrow can be harvested, frozen and reinfused after the cancer cells have been elimi-nated by radiation and chemotherapy. But instead of using marrow it is now possible to isolate cells called 'stem cells', which can be reinfused to repopulate the haematopoietic system after cancer-cell elimination. Stem cells can divide rapidly and give rise to all the dif-ferent types of blood cells. They are thus called 'pluripotent' and are progenitors of all blood cells. Stem-cell technology, tissue matching and the use of less toxic and more specific agents to remove cancer cells have led to the success of bone-marrow transplantation in many haematological malignancies. In recognition of his work on bone-marrow transplantation, Donnall Thomas received the Nobel Prize for Medicine in 1990. He shared it with another pioneer of transplantation medicine, Joseph Murray.

*

Murray, a surgeon from Boston, was also a keen follower of Medawar's work. In 1954, he performed the first successful kidney transplant in a human. The patient, Richard Herrick, had presented with kidney failure. The boy was irritable, and his vision was impaired. He was also prone to convulsions. His blood pressure was found to be very high, his urine contained copious amounts of protein and there were signs of congestive heart failure. On the positive side, Richard had an identical twin, Ronald. The decision was made to carry out a kidney transplant. Before the operation, cross-skin grafting was used to confirm that the twins were genetically identical. The transplant was a success, a world first.[8] Later, Murray also performed the first transplant between a pair of nonidentical siblings.

Murray first met Medawar when the latter visited Boston; they got along well, and Medawar was treated to a ward round at the Peter Bent Brigham Hospital, where he was introduced to several of Murray's renal-transplant patients.

But Medawar's methods of inducing tolerance could not be applied to solve the biggest problem of human transplantation: rejection. In fact, after one of his lectures in America, Peter Medawar was approached by a student who asked him if his foetal-mouse experiments on acquired tolerance had any clinical relevance. 'None whatsoever,' Medawar replied.[9]

However, in his Nobel Lecture in 1990 Joseph Murray paid homage to Medawar, citing his original work carried out at the Burns Unit in Glasgow:

Skin grafts from family members seemed to survive longer than those from unrelated donors. But even after observing hundreds of skin allografts, one could not be certain about their survival time. One certainty was established when Dr J. B. Brown of St Louis in 1937 achieved permanent survival of skin grafts exchanged between MZ [monozygous] twins.

This single observation, although restricted in application, was the only ray of light in the problem of tissue and organ replacement until Gibson and Medawar demonstrated that a second allograft from the same donor was rejected more rapidly than the first. This clear description of the 'second set' phenomenon established that the rejection process was not immutable; instead it implied an allergic or immunological process which potentially might be manipulated.[10]

Compared to the arsenal of immunosuppressive drugs available today, Joseph Murray's 'manipulations' were rather crude. He used sublethal body irradiation and highly toxic anti-cancer drugs, which, apart from causing profound immunosuppression, had terrible side effects. Cortisone was also being used at the time with some success.

But there was hope on the horizon. In 1959, Robert Schwartz and William Dameshek discovered the first immunosuppressive drug, 6-mercaptopurine, an analogue of adenine, one of the purine bases found in DNA. In their original paper, published in 1959, the authors showed a significant prolongation of the graft rejection time in animals treated with the drug. This demonstration of immunosuppression in rats heralded the era of immunosuppressive drug treatment.[11] A related drug, azathioprine, which was found to be less toxic and more effective than 6-mercaptopurine, was also used in organ transplantation.[ii12]

Medawar and his team also studied the role of anti-lymphocyte antibody in transplantation. By depleting lymphocytes, anti-lymphocyte antibody was found to abrogate anti-graft responses in mouse and human transplantation.[13] It is still used in some protocols to delete graft-attacking lymphocytes; the same property has been exploited in the treatment of some autoimmune conditions such as multiple sclerosis.

Emboldened by the availability of effective immunosuppressive drugs, surgeons attempted to transplant more organs. In 1967, Thomas E. Starzl carried out the first successful liver transplant, and in the same year Christiaan Barnard from South Africa performed the first heart transplantation.

We now have a clearer picture of the immune responses that lead to graft rejection. The most common type of rejection is mediated by lymphocytes. Pre-formed antibodies – capable of reacting against donor antigens – found in the recipient mediate a type of rapid rejection, but this is extremely rare due to modern tissue-matching techniques.[iii]

In September 1987, Peter Medawar, whose discoveries had inspired a generation of transplantation biologists and surgeons, suffered another stroke and died eight days later. He was 72.

i   Thomas and his colleagues closely followed the work of Jean Dausset and others (we will come across them in Chapter 27) and successfully carried out their first bone-marrow transplant on a patient with advanced leukaemia, using HLA matched marrow from a sibling.

ii   We can identify three types of rejection: 'hyperacute', 'acute' and 'chronic'. Hyperacute rejection happens within minutes to days following transplantation, and is due to the recipient having pre-existing antibodies against graft antigens. These antibodies are there because the recipient had previously been exposed to foreign human leucocyte antigens (HLA) and had become sensitized to them. This can happen if the recipient had had blood transfusions, multiple pregnancies or a prior transplant. These antibodies, typically directed against major histocompatibility complex

Mary Wortley Montagu
(1689–1762)

Edward Jenner
(1749–1823)

Louis Pasteur
(1822–95)

Élie Metchnikoff
(1845–1916)

Paul Ehrlich
(1854–1915)

Robert Koch
(1843–1910)

Jules Bordet
(1870–1961)

Emil von Behring
(1854–1917)

Karl Landsteiner
(1868–1943)

Peter Medawar
(1915–87)

David Vetter
(1971–84)

Clemens von Pirquet
(1874–1929)

Ralph Steinman
(1943–2011)

Jonas Salk
(1914–95)

William Coley
(1862–1936)

(MHC) class I antigens on the vascular endothelium of the transplant, can also be generated against blood-group antigens, hence the absolute requirement for ABO compatibility between donor and host. The binding of such antibodies to antigen leads to the activation of complement, which causes destruction of the transplanted tissue and thrombosis in the capillaries. These, by preventing vascularization, will cause the death of the graft. Kidneys are most susceptible to hyperacute rejection. Testing the recipient's blood for the presence of any antibodies that might react with the transplant prior to carrying out the operation has greatly reduced the incidence of hyperacute rejection.

Acute rejection, in contrast, is the most common type of rejection seen in the clinical setting. This usually happens within the first six months of transplantation and primarily involves an anti-graft T-cell response. We know this because allografts given to 'nude' mice and rats, which congenitally lack $CD4^+$ T cells, are tolerated without the need for immunosuppression.

$CD4^+$ T helper cells are responsible for orchestrating the process of graft rejection, and they do this by recruiting a range of effector cells: macrophages, $CD8^+$ cytotoxic T cells, natural killer cells and B cells that, by various mechanisms, cause destruction of the transplant.

The rejection process can be divided into two phases, which reflect the generation of a typical immune response: an 'afferent' phase, when sensitization occurs, and an 'efferent' phase, when several immune mechanisms lead to graft rejection. In the afferent phase, MHC molecules found on donor cells, like dendritic cells, are recognized by the recipient's $CD4^+$ T cells, a process called 'allorecognition'. This can happen in two ways. In the first kind of allorecognition, donor MHC is recognized as an intact molecule on the surface of donor antigen-presenting cells (APC) by the recipient's T cells. This is called direct allorecognition. In the second type of allorecognition, a peptide fragment from the donor MHC molecule is presented by a recipient APC to the recipient T cell. This is called indirect allorecognition. There is also a third type called semi-direct allorecognition.

Once the T cell has been presented with donor antigen, it gets activated and can begin the process of sending out effector cells to destroy the foreign tissue. Initially, a non-specific inflammatory response is generated, and this causes an increase in MHC class II antigen expression by the graft, which then leads to more donor antigens being presented to T cells. This initial response can also promote the shedding of donor MHC molecules that can expand the immune response by engaging the indirect allorecognition pathway. In the effector phase, specific $CD4^+$ T cells initiate macrophage-mediated delayed type hypersensitivity (DTH) responses, help B cells to make antibodies against graft antigens and activate cytotoxic T cells that can destroy graft cells either through perforin and granzymes or by inducing apoptosis. Several cytokines, also produced by these activated T helper cells, can cause, among other effects, a macrophage infiltration of the allograft. All these downstream effects then cause the inevitable destruction of the transplant.

Sometimes, a type of rejection called 'chronic rejection' can happen months, even years after the transplantation. The mechanisms of chronic rejection could be the same ones that mediate acute rejection, but could also involve other mechanisms, such as infection of the graft with viruses like cytomegalovirus (CMV), organ fibrosis, arteriosclerosis of graft vessels and toxicity caused by anti-rejection drugs.

iii In the 1980s, a new immunosuppressive drug, produced by a fungus, was discovered. Cyclosporine, by blocking activation of T cells by foreign antigens, prevented T-cell responses against transplants. Cyclosporine and newer drugs like tacrolimus were also less toxic.

Chapter 20

THE MYSTERIOUS LYMPHOCYTE

Bergen-Belsen, Germany, April 1945. James Gowans, a 21-year-old medical student from King's College Hospital, London, was in his designated hut feeding emaciated prisoners with a protein drink called 'hydrolysate'. He was one of the 97 medical students from five London medical schools who had volunteered to help save starving Dutch children recently liberated from Nazi occupation. While en route to the Netherlands, they were diverted to Bergen-Belsen.

'Belsen is a place beyond all imagination,' wrote one student. 'It was originated for death by slow starvation. One's first job is to find the dead and get them removed. Our job is keeping alive those in the huts until they can be got to the hospitals.'[1]

According to records, the arrival of the medical students made a significant difference to the death rates in the camp.

After the war, Gowans obtained his medical degree from the University of London, and in 1953 began working towards a doctoral degree at the Sir William Dunn School of Pathology in Oxford under the supervision of Howard Florey.

Meanwhile the lymphocyte had entered the arena. The stage was set to explore further the role of this mysterious cell in skin- and organ-graft rejection, delayed-type hypersensitivity, tumour rejection and all *specific* immune responses in general. It is important to remember that we are talking about the specific immune system: Metchnikoff's phagocytes, like macrophages and neutrophils, were part of the innate or non-specific immune system.

It was Florey who suggested that Gowans should investigate the lymphocyte. Lymphocytes, Florey remarked, had already blunted the wits of several of his colleagues. And he could see no reason why Gowans should be spared a similar fate.[2]

At the time, the small lymphocyte was considered to be an end cell with a very short lifespan and no discernible importance. By removing some of these cells from the thoracic duct – the largest lymphatic vessel in the body, which contained these cells in large numbers – and labelling them with radioisotopes, Gowans showed that they were in fact not short-lived but recirculating back and forth from the blood into lymph, entering then leaving the swellings found along the lymph vessels: the lymph nodes.[3] But why would these cells recirculate? No one could answer that question.

It was also found that depleting lymphocytes by draining the thoracic duct in rats impaired their ability to mount antibody primary responses to injections of tetanus toxoid or sheep red blood cells. The rats were also unable to reject skin grafts. Reinjection of thoracic-duct lymphocytes from another syngeneic – that is, a pure-bred rat – restored all these immune responses.[4]

In another experiment, an immunologically virgin rat – one that had not had the pleasure of interacting with a certain antigen – when inoculated with lymphocytes from a rat that had already mounted a primary response to that specific antigen, mounted a secondary response when exposed to the same antigen.[5] This suggested that the lymphocytes somehow carried a 'memory' of the first contact and thus generated the secondary immune response. Suddenly all the features of *specific* immune responses known up until that time – heightened responses to second exposure, the mechanism underlying vaccination, hypersensitivity and graft rejection – appeared to be carried out by the small lymphocytes. This work also explained why lymphocytes recirculated. They were on patrol looking for antigens. Seek and destroy.

In a review of this period, James Gowans cites his mentor

Howard Florey: "'The complete ignorance of the function of this cell is one of the most humiliating and disgraceful gaps in all medical knowledge... Literally nothing of importance... is known regarding the potentialities of these cells." This was the situation in 1953.'[6]

Encouraged by Peter Medawar, Gowans then turned his attention to the role of the lymphocyte in acquired tolerance. One ingenious experiment demonstrated that tolerance was mediated by the small lymphocyte. In the experiment, a non-tolerant strain-A mouse given strain-A lymphocytes taken from a mouse made tolerant to strain-CBA tissues was able to accept grafts from a CBA mouse. But when a non-tolerant strain-A mouse was injected with cells from a strain-A mouse that had *not* been made tolerant and then given CBA graft, the graft was rejected. In this case, a graft-versus-host reaction also occurred: donor lymphocytes attacking host tissue.[7] The conclusion was clear: to confer tolerance, the injected cells had to come from an animal that had been made tolerant. And the key cell type mediating this was the lymphocyte!

Gowans's findings unleashed a torrent of research that confirmed the central role of the lymphocyte in immune responses. These cells are found in lymph, the tissue fluid that flows inside the extensive system of vessels that forms the lymphatic system. Where did this lymph come from? Lymph is derived from blood that seeps out of capillaries, the tiny blood vessels that carry oxygen and nutrients to all tissues. These capillaries sprout from small arteries and, due to the increased pressure at the arterial end of the capillary bed, tissue fluid flows out of the vessels. But because proteins in the blood are retained within the capillaries, osmotic pressure draws most of the fluid back in at the venous end of the capillary bed, with any excess fluid ending up in the lymphatic vessels as lymph. Inside the lymphatics the lymphocytes circulate, like policemen on the beat, looking for antigen. These thin, transparent vessels are punctuated by swellings called lymph nodes, which can be easily felt in the armpits, neck or groin when they become swollen. Immune responses against antigens found in the lymph are mounted inside these lymph nodes: police stations where the weight of the law is unleashed on the criminals captured by the cops on the beat. Extending the analogy, a swollen node can be thought of as a police station chock-full of criminals being processed by eager law enforcers!

All the lymphatic vessels unite to form two large vessels, the thoracic duct and the right lymphatic duct, which end up being plumbed into the circulatory system via the subclavian veins. So lymph that came out of the blood goes back into the blood: the circulatory and lymphatic systems are intimately connected.

James Gowan died in April 2020. Colonel E. E. Vella, one of the senior army doctors who had been at Belsen wrote:

The arrival of 97 medical students from the London Teaching Hospitals proved the greatest help and with their advent the death rate, which in the earliest days had been 500 per day, began to drop appreciably. With their knowledge and enthusiasm, we were able to exercise much better supervision in each hut; they worked splendidly, and I cannot speak highly enough of their efforts.[8]

Chapter 21

THE BOY IN THE BUBBLE

What happens when the immune system is deficient in one or more of its components? What happens if antibodies cannot be generated or lymphocytes fail to mount immune responses? Up until the 1950s no one had any idea, because no case of immunodeficiency had been documented. That crucial discovery – that there were indeed individuals with deficient immune systems – came from America.

In 1952, Colonel Ogden Bruton, an American army paediatrician, described an eight-year-old male who presented with an unusual constellation of infections. The boy suffered from chills, a fever of 38.1°C and left-knee pain. History did not reveal any significant illnesses; his mother had had an uneventful pregnancy; the delivery was normal, and the boy's developmental history was also unremarkable. The attending physician, suspecting rheumatic fever, prescribed aspirin and discharged the boy. However, there was no improvement, and within two hours his temperature had spiked up to 38.8°C, necessitating a return to the hospital. This time the boy was admitted. He was treated with penicillin for 28 days and discharged. But after a few weeks the boy was back, this time with an upper-respiratory-tract infection which had turned into pneumonia: his temperature had now gone up to 40°C. He was given sulphonamides and recovered some five days later. But within a week he returned, complaining of a painful jaw swelling. Again, he responded to treatment and was discharged, but a few weeks later was back in hospital, vomiting from a violent gastrointestinal upset and running a high fever. Once again he was treated with sulphonamides and sent home, only to return just a week later with fever and vomiting, which lasted for two weeks. He then developed severe middle-ear infections in both ears. Treated with penicillin in oil and beeswax, he made a good recovery, but his left eardrum had required incision. However, blood samples taken at the time had grown a bacterium, the pneumococcus that had caused the pneumonia. Alarmingly, the boy had a blood infection: he was septicaemic. Two months later he again developed pneumococcal middle-ear infections, but again responded to sulpha drugs.

This waxing and waning continued for several more months. Frequent infections were not uncommon in children, but Ogden Burton was troubled by the fact that the boy had suffered from clinical sepsis, a serious bloodstream infection, on six separate occasions with pneumococci isolated in three out of the six. Prophylactic antibiotic therapy was failing to stop these recurrent episodes and the boy was admitted for further investigation.

Because the initial episodes of sepsis involved the same type of pneumococcus, a vaccine was prepared from bacteria isolated from the patient. This was given over five months, but the boy failed to make any antibodies. As more types of pneumococci were being isolated, a vaccine containing six different types isolated from the boy was prepared and administered over a period of seven months. But, again, no antibodies were detected. Subsequent investigations found that the boy had not made any antibodies in response to his routine childhood immunizations for diphtheria, tetanus or typhoid.

As we shall see in Chapter 25, antibodies are found in the gamma globulin fraction of serum, and when the boy's serum was examined it was found to be negative for gamma globulin. Bruton then decided to give the boy human-serum-immunoglobulin injections; six days later repeat serum analysis revealed a large amount of gamma globulin: the boy now had a gamma globulin fraction. For the next 14 months the boy was given monthly injections of gamma globulin as the levels were found to decrease gradually over time, disappearing in six months. But the injections stopped the episodes of sepsis: Bruton had found an answer to the repeated infections. The boy was treated regularly with antibiotics and immunoglobulin replacement: he had a normal childhood, finished school and entered college.[1]

After Bruton's publication many physicians came forward to report similar cases, and analysis of these revealed this to be an X-linked disease: it only affected boys. Interestingly, while Bruton's boys suffered from frequent bacterial infections they did not unduly suffer from viral infections, and their delayed-type hypersensitivity responses were intact. The immunodeficiency appeared to involve only the antibody response.

*

Also in the 1950s, Swiss paediatricians Eduard Glanzmann and Paul Riniker described another congenital form of immunodeficiency. They found unusual patterns of infection in siblings from certain families. These children suffered from terrible yeast infections, the infection often spreading from the throat into the larynx, oesophagus, stomach and bowels. Laboratory investigations revealed that the children had reduced levels of lymphocytes, and, like Bruton's boys, no antibodies.[2] In 1961 reports appeared of a fatal disease involving agammaglobulinaemia – no antibodies – and lymphopenia – hardly any lymphocytes. These diseases, collectively termed 'Swiss-type immunodeficiency', involved both the antibody and cellular compartments of the immune system.

More case reports began to appear as physicians became increasingly vigilant for patients who presented with signs and symptoms suggestive of immunodeficiency conditions. A familial disease that showed an autosomal recessive inheritance pattern was described where patients had agammaglobulinaemia, low lymphocyte counts and small, underdeveloped thymus glands. In contrast to Bruton's X-linked agam-maglobulinaemic patients, these patients were all unusually susceptible to fungal, viral and some bacterial infections. They also couldn't make antibodies or mount the delayed-type hypersensitivity reaction, again indicating deficient antibody and cellular responses.

But, in the 1970s, the story of one boy's battle with a severe immunodeficiency spread throughout the world, bringing sharply into focus the devastating consequences of living without an immune system.

*

David Phillip Vetter was born in 1971. His parents, David Joseph Vetter, Jr, and Carol Ann Vetter, had lost their first son, seven months after he was born, to a rare type of X-linked immunodeficiency disease, SCID, which stands for 'severe combined immunodeficiency disease'. They had another child, a healthy daughter, Katherine, who was born four years before David. When she was five months pregnant, Carol had amniocentesis, which revealed that she was carrying a baby boy. The doctors informed her that there was a 50 per cent chance that her baby would be profoundly immunodeficient, but the Vetters decided to go ahead with the pregnancy. As soon as he was born, David was placed in a sterile incubator. Soon, tests confirmed the parents' worst fears: David had the same illness as his brother. He was deficient in both the antibody and cellular compartments. He had SCID.

The only option available to the doctors was to keep David in isolation: in a sterile environment, a plastic bubble, until a bone-marrow transplant from a compatible donor could be performed. Water, air, food and clothing had to be sterilized before being passed into the bubble and David was handled by his parents and medical staff through gloves attached to ports made in the walls of the bubble. He was given a normal education, and the bubble contained a television and a playroom. But the constant hum of compressors pumping sterile air into the bubble made it very noisy and David was often irritable and visibly unhappy; when he was four David poked holes in the chamber with a butterfly syringe that had been left inside.

In 1977, NASA developed a special suit, connected to the bubble by an eight-foot tube, that allowed David to go outside. But the boy was showing signs of frustration and it became clear to all concerned that the bubble was not a permanent solution.

In 1983, having failed to find a perfect match, the Vetters agreed for David to receive a bone-marrow transplant from his sister Katherine: doctors felt that he had a significant chance of success as bone-marrow transplantation had progressed to a point where partial matches often worked. The operation, carried out in a sterile theatre, went smoothly, and initially David appeared to be doing well. But within a few weeks things began to go horribly wrong. David was taken to hospital suffering from what was thought to be graft-versus-host disease. He had an elevated temperature and abdominal pain, was vomiting and began losing weight. Just four months after the transplant, David Vetter, the boy who had spent his entire life in a bubble, lost his battle. He was just 12 years old.

An autopsy revealed that David had not died of graft-versus-host disease: no traces of Katherine's bone marrow were found – the graft had failed. The cause of death was Burkitt's lymphoma, a cancer of lymphocytes which had spread throughout David's body.[i] So where did the lymphoma come from? Further analysis revealed that the donor bone marrow contained traces of Epstein–Barr virus (EBV). Was this the cause of David's cancer? Did the virus cause the lymphoma? It was known that EBV caused glandular fever, mononucleosis, and, unbeknown to any of the doctors, Katherine had suffered from mononucleosis early in her life. David had acquired the infection from infected bone-marrow cells donated by his sister.[ii] Eventually the connection was made, the small silver lining in the tragic case of David Vetter. For the first time, a virus was shown to cause a human cancer.[3]

David Vetter suffered from X-linked severe combined immuno-deficiency, X-linked SCID. Patients with SCID, like those described by Glanzmann and Riniker, are prone to recurrent infections, show slower growth and normally do not live beyond infancy. We now know that this condition is caused by a mutation in a gene found on the X chromosome. Other primary immunodeficiency diseases, such as Bruton's agammaglobulinaemia, are caused by mutations in different genes.[iii]

i The lymphoma B cells were David's: his type of immunodeficiency did not have a deficiency in the absolute count of B cells; rather the problem was in the lack of functioning B cells. David's B cells were infected with Epstein–Barr virus (EBV), which caused their transformation into a malignant phenotype.

ii Ordinarily such a viral infection would have been neutralized by T cells, but David did not have any effective T cells to fight the infection.

iii David Vetter suffered from X-linked SCID. The deficiency was due to a mutation of the gene coding for the common gamma chain, which is a constituent of several cytokine receptors, such as IL-2R, IL-4R, IL-7R, IL-9R, IL-15R, and IL-21R. Without a functional gamma chain, theses cytokine receptors cannot function, and the inability to bind cytokines results in loss of cytokine function. The common gamma chain gene is found on the X chromosome; therefore, this variant of SCID is X-linked. Approximately 50 per cent of all patients with SCID have this gene defect. Why does loss of cytokine function cause immunodeficiency? The cytokine IL-2 is required for the proliferation of T, B and NK cells; therefore, if the IL-2R doesn't function correctly, lymphocyte prolifera-tion signals are impaired. Loss of IL-4R function also prevents B cells to class-switch so IgG responses cannot be mounted. By being unable to stop unwanted T-cell apoptosis, loss of IL-7R function leads to an incapability of T-cell selection to occur in the thymus. Additionally, loss of IL-7R function is associated with the lack of a functional T-cell receptor (TcR). Loss of IL-15R function can prevent the development of NK cells.

In the second most common form of SCID, a key enzyme, adenosine deaminase (ADA), is defective. ADA is necessary for the breakdown of purines, and lack of functioning ADA causes deoxyadenosine to accumulate inside cells, which in turn leads to the accumulation of deoxyadenosine triphosphate (dATP), which by inhibiting the activity of ribonucleotide reductase – the enzyme that reduces ribonucleotides to generate deoxyribonucleotides – causes a reduction of DNA synthesis. As T- and B-lymphocyte proliferation, which requires a steady supply of DNA, is an essential requirement for specific immune responses, the lack of ADA causes profound immunodeficiency.

In 1990, a four-year-old girl, Ashanthi DeSilva, who had ADA deficiency, became the first person in the world to be treated by gene therapy. This involved the removal of some of her white blood cells, which were infected by a virus carrying the normal version of the ADA gene. These cells, with the normal ADA gene incorporated into their genome, were then injected back into Ashanti. To date she is well.

Following on from this, a similar approach has been used in X-linked SCID where a normal gene has been introduced into stem cells using a viral vector delivery system. Twenty SCID babies have been treated by this method, and 18 of them are still alive. However, the procedure is not risk-free, and some children developed leukaemia because of the treatment.

A T-cell deficiency is always found in SCID, and antibody production is severely impaired, even when mature B cells are present, because of the lack of T-cell help.

The gene defect in Bruton's agammaglobulinaemia has been mapped to a gene on the X chromosome that codes for an enzyme called Bruton tyrosine kinase (Btk). The Btk enzyme is required for the proliferation and differentiation of B lymphocytes. Therefore, a mutation in the Btk gene can lead to a deficiency of a functional enzyme required for the normal division and maturation of B cells; this will therefore cause a disruption in the production of antibodies.

Chapter 22
THE STRANGE ORGANS OF BIRDS AND MICE

Manchester, England, 1951. An eight-year-old boy had his spleen removed because of a rare blood disorder: hereditary spherocytosis. Defective red blood cells were being destroyed by macrophages, leaving the boy anaemic. The destruction took place in the spleen as the fragile red blood cells passed through the organ. Removal of the spleen was a last resort. Twenty-six years later, when he was 33, he was back at the hospital, admitted with a sudden fever and extreme tiredness. According to one of the doctors who tried to treat him, 'He rapidly went cold and blue, lost consciousness, and died.'[1] The cause of death was an overwhelming infection.

Doctors at this Manchester hospital described two other cases in which infection had developed some 14 and 25 years after splenectomy, causing the death of one of the patients. This association between splenectomy and infection had only been recognized in the late 1960s, and so none of the patients were on prophylactic antibiotics. The illness had come out of the blue, without warning. It was more common in children but could happen at any age. The risk was highest in the months and years after removal of the spleen, but never disappeared: there was a case of infection 45 years after splenectomy. Usually the reason for a splenectomy is a traumatic event such as a road traffic accident in which the spleen is damaged, and the soft, dark-red organ tucked under the left ribcage must be removed to stem the internal bleeding. Patients without a spleen require lifelong penicillin prophylaxis. The reason that patients without a spleen developed overwhelming infection that sometimes led to a fatal sepsis decades after splenectomy was not known at the time.

*

In 1952, Bruce Glick, a graduate student from Ohio State University, began studying a mysterious organ first described in 1621 by the Italian anatomist Hieronymus Fabricius. This small outpouching at the tail end of the digestive system, found only in birds, was called the 'bursa of Fabricius', and its function was unknown. Attempting to discern its function, Glick removed the bursa from chickens, but the birds did not appear to be affected. Timothy Chang, a fellow doctoral student, required some birds to generate antibodies against salmonella, and Glick offered him the bursectomized animals. To Chang's surprise and annoyance – he really needed the antibodies – the chickens failed to produce antibodies against salmonella. Had the bursectomy – the removal of the bursa – somehow caused the failure of antibody production in the chicken?

Intrigued, Glick and Chang carried out further experiments that established the fundamental role of the bursa of Fabricius in the production of the antibody response in birds. But they also found that the bursectomized birds could overcome viral infections and reject skin grafts, confirming that cellular immunity was left unimpaired. As we have seen before, immunity to viral infections and graft rejection are mainly mediated by immune lymphocytes. Believing that they had stumbled across a significant discovery, Glick and Chang wrote up their findings and submitted the paper to the prestigious journal *Science*. The editors of *Science*, however, disagreed and, finding the study uninteresting, rejected the paper. Glick and Chang did manage to get their work published, in 1955, but only in the obscure journal *Poultry Science*.[2]

There was another organ found in vertebrates whose function remained unknown. This was the thymus, a strange, bilobed organ found overlying the heart in the middle of the chest. Prominent in young animals, it mysteriously shrank and disappeared with maturity. The function of the thymus was discovered by Jacques Francis Miller.

*

Miller, who is sometimes described as 'the last person in the world to discover the function of a human organ', was born in 1931, in Nice, France. In 1941, at the height of the Second World War, his family migrated to Sydney, Australia, where Jacques grew up and completed his education, obtaining a medical degree from the University of Sydney.

He was studying certain lymphomas in mice when he made the discovery. He found that the lymphomas first occurred in the thymus gland and then spread to other parts of the body. When he performed a thymectomy – removed the thymus – in adult mice he found that the lymphomas were still occurring elsewhere. Believing that the cancer cells had already spread from the thymus into the periphery, he decided to remove the gland early in life and carried out thymectomies in newborn mice. To Miller's surprise, the thymectomized mice failed to thrive and became runts; they were also unable to reject grafts. Miller wrote that he had mice with four or five different types of skin on their backs, of all colours.[3] The mice failed to elicit the delayed-type hypersensitivity response; interestingly, their ability to make antibodies against sheep red blood cells was also impaired. Autopsy revealed that the mice had relatively few lymphocytes. There was only one explanation: the thymus gland was essential for cell-mediated responses like skin-graft rejection and for the generation of lymphocytes. Interestingly, some antibody responses also appeared to require the input of the thymus.[4]

By now the importance of the lymphocyte in immunity had been established. Lymphocytes mediated both antibody responses and cellular responses such as resistance to viral infection and the rejection of skin grafts. The work of Glick and Miller suggested that antibody and cell-mediated immune responses were carried out by two distinct types of lymphocytes. In birds a specialized organ – the bursa of Fabricius – presumably made the antibody-producing lymphocytes, whereas it appeared that the thymus gland produced the lymphocytes mediating cellular immunity and played a role in antibody production. But exactly how these events happened was a mystery.

Some of the seminal research that helped confirm these suppositions was carried out by the American physician–scientists Robert Alan Good and Max Cooper.

*

Good was born in 1922, in Crosby, Minnesota. At the age of six, young Robert watched his father losing his battle against testicular cancer. As a result, he wished to find a cure for cancer, and so decided to pursue a career in medicine. In 1941, he was offered a place at the University of Minnesota School of Medicine but unexpectedly fell ill with Guillain–Barré syndrome, a disease in which autoantibodies attack the peripheral nervous system. Robert was left paralysed, and the dean of the School of Medicine decided that because of this severe disability he would not be able to honour the offered place. Robert appealed: he might be physically impaired, he said, but his intellectual capabilities were intact, and he deserved his place at medical school. He won the appeal and took up the place. It wasn't easy, and his mother had to push his wheelchair into his lectures. But Good excelled, and in 1944 graduated with honours, with both an MD and a PhD, completed in just three years. He chose to specialize in paediatrics, and completed a residency programme one year later.

In 1953, Good came across a 54-year-old male patient who had been suffering from repeated bouts of pneumonia over several years.

He also had a thymoma, a benign tumour of the thymus gland. Laboratory tests revealed that the man had a reduced ability to make antibodies and mount cellular immune responses. Removal of the thymoma did not correct these deficiencies. After finding similar cases in which patients with thymic tumours also had immuno-deficiency in both the humoral and cellular compartments, Good decided to investigate the role of the thymus in immunity.

Then, in 1963, a young paediatrician called Max Dale Cooper joined Good's laboratory at the University of Minnesota. Familiar with Jacques Miller's work, Cooper also wanted to probe the role of the thymus gland in immunity. In a recent article in *Nature* reviewing this period, the authors commented that 'at that time, antibodies were thought to derive only from lymphocytes... But immunologists did not know whether and how the lymphocytes that produced antibodies and those involved in graft rejection were related.'[5]

In his work as a paediatric immunologist, Max Cooper observed that children with an X-linked immunodeficiency disease called Wiskott–Aldrich syndrome often suffered from severe herpesvirus infections despite having elevated levels of antibodies. In contrast, as we have seen before, boys with Bruton's agammaglobulinaemia could control viral infections even though they were unable to pro-duce antibodies. Then, perhaps serendipitously, Max Cooper came across Glick and Chang's *Poultry Science* paper.

These two observations – of immune-deficient children lack-ing in humoral or cellular immune compartments and of Glick's bursectomy experiments – inspired Good and Cooper to carry out experiments to prove the existence of two lineages of lym-phocytes responsible for humoral and cellular immune responses, respectively.

Their groundbreaking experiment involved two batches of newly hatched chickens. Good and Cooper removed the thymus in one batch and the bursa of Fabricius in another. They found that the chicken without a bursa, when challenged with the protein bovine albumin or the bacterium *Brucella abortus*, did not produce antibodies. They were in effect like the patients suffering from Bruton's disease: they were agammaglobulinaemic. In contrast, the chicks without thymuses could not reject skin grafts, mount delayed-type hypersensitivity or show graft-versus-host responses. In these animals the antibody responses to some antigens were also found to be impaired, even though their total immunoglobulin levels were quite high. This seminal work was published in *Nature* in January 1965.[6]

But an important question now emerged. What was the bursal equivalent in mammals? Was there one? Which organ was responsible for making antibodies in man? Max Cooper was often ribbed by colleagues, who would ask him, 'Where is the bursa equivalent this year, Max?'[7]

But the organ dichotomy of antibody and cellular immunity wasn't absolute. Jacques Miller's work and the experiments of Max Cooper and Robert Good suggested that the thymus was also involved in antibody production.

*

In 1966, Henry Claman, an American physician and immunologist, carried out a series of experiments that attempted to solve this puzzle. These involved subjecting inbred mice to a dose of lethal irradiation and then reconstituting them with thymic cells (thymocytes), bone-marrow cells or both, taken from an unirradiated syngeneic mouse. So one group of irradiated mice received thymic cells, another group bone-marrow cells and the third group both types of cell. All groups of mice were then challenged with sheep red blood cells. Claman then counted the number of cells that were making antibodies to sheep red cells in these mice using a new plaque-forming cell assay that quantified antibody-forming cells. He found that mice that were given syngeneic thymocytes or bone-marrow cells on their own failed to produce antibody-forming cells, whereas mice that were given *both* bone-marrow and thymic cells showed a hundred-fold increase in the numbers of antibody-forming cells. These experiments showed that both thymic-derived and bone-marrow-derived cells were required to generate antibodies against sheep red blood cells. This was also the first demonstration of synergy: cell-to-cell cooperation in immune responses.[8]

But which cell in the mix was making the antibodies, and which cell was acting as the 'helper'? This question was answered by an experiment conducted in 1968 by Jacques Miller and Graham Mitchell, working at the Walter and Eliza Hall Institute in Melbourne. This experiment is complex, so it might be an idea to have a cup of coffee before diving in.

Two strains of mice were used in this experiment: strain Y and strain XY, a cross-bred mouse. A strain-Y mouse was given a lethal dose of X-rays, and a syngeneic cell mixture containing bone-marrow cells from strain Y and thymic cells from strain XY was injected into these animals. When the mouse was then challenged with sheep red blood cells, the strain-Y animals produced antibody. But which cell was making the antibody: the Y bone-marrow cell or the XY thymic cell? Warning! Now it's about to get a bit complicated.

To answer that question, Miller and Mitchell needed to remove one of these cell populations – bone-marrow or thymic cells – from the cell mix and repeat the experiment. They used antisera raised against lymphocytes from strain Y and strain XY mice to remove the respective cell populations. So, adding anti-Y antisera, containing antibodies against strain Y lymphocytes, along with complement, to the cell mix would cause the Y lymphocytes to lyse. Similarly, anti-X antisera would lyse strain-X cells in the mix. Therefore, by using these antisera Miller and Mitchell were able to remove the Y bone-marrow cells and XY thymic cells from the cell mix. When these antibody-treated cell-depleted mixtures were injected into two groups of irradiated strain-Y mice and challenged with sheep red cells, the animals that received cell mixtures pre-treated with anti-X made large amounts of antibody, whereas those mice given anti-Y pre-treated cells did not (Fig. 11). This suggested that the Y bone-marrow cells were responsible for antibody production and the XY thymic cells were acting as the helper.[9]

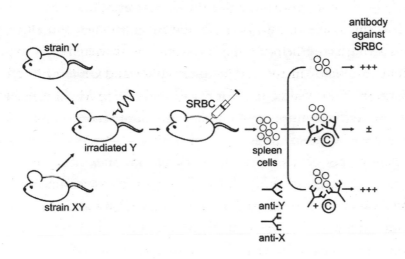

Figure 11. Experiments demonstrating the origins of antibody-producing and helper-cell populations.

In 1969, Angelo DiGeorge, an Italian American paediatric endo-crinologist based in Philadelphia, came across Max Cooper's paper. Four years earlier, DiGeorge and his colleagues had encountered an infant with congenital hypoparathyroidism: lacking a parathyroid gland (found within the thymus), which controls calcium balance in the body. The chest X-ray did not show the expected thymic shadow, and this absence of a thymus gland prompted DiGeorge to question whether the infant, like Max Cooper's thymectomized chicken, also had defective cellular immunity. He carried out a skin graft and found that that was indeed the case: the infant couldn't reject the graft. The infant was also runted, despite adequate control of calcium levels.[10]

DiGeorge found more cases of children lacking a thymus and parathyroid glands, and they too had defective cellular immunity. These infants suffered from persistent candidiasis and could not mount delayed-type hypersensitivity responses. These congenitally athymic infants were indeed like the thymectomized birds described in Cooper's paper. But the total lymphocyte count, the antibody-producing cell numbers in lymph nodes and the serum immuno-globulin levels were all normal in these children. These observations led to the characterization of a new primary immunodeficiency disease: DiGeorge syndrome.

The main abnormality in DiGeorge syndrome is a loss of thymic tissue. This loss is usually not total, and a profound decrease in thymic-derived lymphocytes and consequent severe immunodefi-ciency are only encountered in a minority of patients. In approxi-mately 90 per cent of cases, part of chromosome 22 is missing. However, despite having the same chromosomal abnormality, patients show a wide variation in their signs and symptoms. These include dysmorphic features such as an underdeveloped chin, low-set ears and cleft lip and palate; cardiac abnormalities, notably affecting the aorta; learning and speech problems, and even neuropsychiatric conditions such as schizophrenia. Many patients also have loss of parathyroid glands with low blood calcium levels, which in profound cases can lead to seizures. DiGeorge patients are also more likely to suffer from autoimmune diseases such as autoimmune haemolytic anaemia, idiopathic thrombocytopenic purpura and autoimmune thyroid disease compared to the normal population, reflecting the importance of the thymus gland in the maintenance of tolerance.

The thymus-derived lymphocytes were called 'T lymphocytes' and the bursal-derived (or bone-marrow-derived, in mammals) lympho-cytes, 'B lymphocytes'. Soon the leading role of the T lymphocyte in all types of specific immune responses became apparent: the T cell was the conductor of the immunological orchestra. It was subsequently found that both types of lymphocytes, B and T, were in fact produced in the bone marrow, the T cell migrating to the thymus to mature and complete its development.

We now know that there are several types of T cell, each with a different function in immune responses. Broadly speaking, three types of T cell can be identified: helper, cytotoxic and regulatory. Some of these T cells, called 'effector T cells', are typically short-lived and carry out immune responses; others, called 'memory cells', remain dormant, giving rise to the heightened secondary responses after a second exposure to antigen. The T helper (Th) cells assist other immune cells in immune responses. These include the maturation of B cells into antibody-producing plasma cells and memory B cells that give rise to the secondary antibody response, the activation of cytotoxic T cells, and, by activating macrophages, enhancement of innate immunity. All T helper cells express the CD4 glycoprotein on their surface. To complicate matters, several subtypes of helper cells have been described, each with distinct functions![i]

Cytotoxic T cells, on the other hand, have the CD8 glycoprotein marker on their surface. Their job is to destroy cells by killing them directly: those infected with unwanted pathogens inside them, tumour cells and, unfortunately for those requiring transplants, allografts.

Similarly, there are effector and memory B cells, the former differentiating into large plasma cells that make antibody and the latter generating secondary antibody responses. Subtypes of B cells have also been described. Additionally, B cells have another important function that we will consider in detail in Chapter 27: they present antigen to T cells.[ii]

We can now consider the organs of immunity. Two types of lymphoid organ have been defined: primary and secondary. The thymus, bone marrow and the bursa of Fabricius in birds are primary lymphoid organs. Immature lymphocytes generated in the bone marrow mature and acquire specific antigen receptors in primary lymphoid organs (Fig. 12).

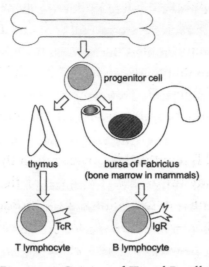

Figure 12. Origins of T and B cells.

These lymphocytes thus become immunocompetent, and when they leave the primary lymphoid organs they are ready for action. They also undergo a selection process in the primary lymphoid organs, where any cells likely to react against self-antigen are deleted. Primary lymphoid organs are like military training schools where 'immature' cadets are carefully selected – the trigger-happy or lazy ones being weeded out – and armed before being sent out to engage with the enemy.

Let's start with the thymus. This gland, made up of lobules, reaches its maximum weight of around 30 grams by the time of puberty. A functional thymus during the embryonic and neonatal period appears to be vital for the generation of effective immune responses, and if the thymus is missing or reduced in mass, immunodeficiency, with a consequent increased susceptibility to infection, occurs. The thymus starts to degenerate soon after puberty, and as we enter old age it is a small, insignificant bit of tissue, mostly made of fat. This natural loss of thymic tissue with ageing could be one reason we become more susceptible to infections and cancers as we get older.

As we saw earlier, the cells that enter and mature in the thymus are called T cells. Once mature, these T cells leave the thymus and make up the peripheral T-cell population that engages pathogens and altered cells.[iii]

The cells that complete their maturation in the bursa in birds, and in bone marrow in mammals, are called B cells. And we now have the answer to the question Max Cooper was asked at annoyingly regular intervals: the bursal equivalent in the human is... the bone marrow!

So, these T and B lymphocytes, armed with their specific receptors, are now ready for action. The basis of their specificity, one of the hallmarks of the specific immune response, is the receptors sticking out of their cell surface like locks that can fit the antigenic keys possessed by pathogens. As we saw in Figure 12, each T or B cell has one specific receptor that binds to a single antigen.

After leaving the primary lymphoid organs, immunocompetent lymphocytes carry out their effector functions in secondary lymphoid organs: engaging antigen and mounting immune responses. The spleen, lymph nodes and diffuse collections of lymphoid tissue found underneath the linings of respiratory, gastrointestinal and urinogenital tracts are secondary lymphoid organs. This is where the cadets, now fully armed soldiers, engage the enemy.

Let us first consider lymph nodes. These small nodules, found along lymphatic vessels, are scattered throughout the body. As such they are ideally placed to 'drain' the body tissues, mounting immune responses against pathogens that have ended up in the lymph. So when a pathogen, having breached the innate barriers, gets into the tissues it is taken by the lymphatic vessels to the local lymph nodes, where it will face the full force of the specific immune system.[iv]

Next, we can look at the spleen. In humans the spleen is found in the upper-left quadrant of the abdomen, lying underneath the ribcage. In a healthy adult it weighs around 150 grams. It deals with blood-borne pathogens, mounting immune responses against any pathogen that has managed to get into the bloodstream. As we saw at the beginning of the chapter, its vital importance in clearing blood infection is exemplified by cases in which the spleen has been removed after being damaged by a traumatic injury. Patients who have had a splenectomy require antibiotics for life.

The spleen has other essential functions: it's a scrapyard or, more accurately, a recycling centre. Old and defective red blood cells are destroyed in the spleen, as are antigen antibody complexes that, if left lying around, can cause unwanted inflammation.[v]

There is a vast collection of secondary lymphoid tissue, clustered beneath epithelial surfaces and collectively called mucosa-associated lymphoid tissue (MALT), which mounts immune responses against antigens that enter the body through mucosal surfaces. Our gastro-intestinal, respiratory and urogenital tracts are all lined by mucosal surfaces, and as such are open to the outside: pathogens can therefore enter the body through this surface. MALT is populated by lymphoid tissue containing antigen-presenting cells (we will come across these cells again in Chapter 27), T and B lymphocytes and plasma cells. These cell clusters in the gastrointestinal system are overlaid by specialized cells in the epithelium called M cells, which sample antigen from the gut lumen before delivering them to the MALT underneath.

i   Specific immune responses begin when antigen-presenting cells such as dendritic cells, macrophages and B cells capture and present foreign antigens to the T helper cell. Usually the antigens are derived from exogenous pathogens, and once internalized the antigenic peptides associate with MHC class II molecules and the MHC–peptide complex gets displayed on the surface of the APC. The T helper cell, through their specific receptors, binds the MHC class II–peptide complex. If co-stimulatory signals are also provided, the T cell becomes activated. Once activated, the T helper cell can help in antibody production, activate cytotoxic T cells and other cells harbouring intracellular pathogens.

To destroy an infected or altered cell, the cytotoxic T cell must bind, through its specific receptor, a foreign or altered antigenic peptide presented to the T cell by an MHC class I molecule. MHC class I molecules are present on all nucleated cells; hence any such cell, if infected or altered, can be destroyed by a cytotoxic T cell.

ii As Ehrlich suggested in his side-chain theory, the B-cell receptor, a surface immunoglobulin molecule, is the actual antibody that is subsequently secreted by the cell. Antigen binds to the B-cell receptor, gets internalized, combines with MHC II and is presented to a T helper cell. The activated T cell, through the production of cytokines, helps the B cell to develop into a plasma cell that can produce copious quantities of specific antibody elicited against that antigen.

iii After the T cells are generated in the bone marrow they migrate to the thymus. The early events in the development of these immature T cells take place in the cortex. Here T-cell receptor gene rearrangement takes place, resulting in each T cell having on its surface a unique receptor, specific to a limited set of peptide: MHC combinations. Thymocyte selection also begins in the cortex. There are two types of selection: 'positive' and 'negative'. In positive selection, those T cells that cannot bind, or bind too weakly, to MHC-self antigen complexes are deleted: if they cannot recognize self MHC they will be useless at detecting infected cells later. T cells that bind too strongly to MHC-presented self-antigens also undergo deletion. This is called 'negative selection'. It is important that any T cell that binds too readily to a self-antigen–MHC complex be eliminated, as these cells can cause autoimmunity if allowed to leave the thymus. In the thymic medulla, T cells undergo additional rounds of negative selection to ensure that any autoreactive T cells are not present in the final population that is allowed to mature. Those cells that survive positive and negative selection differentiate into either CD4$^+$ T helper or CD8$^+$ T cytotoxic cells: if a T-cell receptor recognizes an antigen–MHC class I complex the cell becomes a CD8 cell; if it recognizes an antigen–MHC class II complex it becomes a CD4 cell.

iv A lymph node consists of a cortex, paracortex and medulla. The cortex contains B cells, macrophages and specialized antigen-presenting cells called 'follicular dendritic cells' (FDCs) clustered within structures called primary follicles. Antigen arriving in the node is taken up by FDCs and presented to B cells that proliferate, forming a germinal centre within the primary follicle and transforming the latter into a larger secondary follicle. Lymph nodes of children with B-cell deficiencies have cortexes lacking primary follicles and germinal centres. The paracortex region is composed mainly of T cells and another antigen-presenting cell called the interdigitating dendritic cell. Antigen presentation and activation of T cells happen in the paracortex; hence this area is called the 'thymus dependent area': lymph nodes removed from neonatally thymectomized mice have empty paracortical regions. The medulla contains mainly antibody-secreting plasma cells. Lymphocytes can enter a lymph node from afferent lymphatics and from the blood through a specialized part of the blood vessel called the 'high endothelial venule' (HEV).

v The spleen has two regions: the 'white pulp' and the 'red pulp'. The immune functions are carried out in the white pulp. The periarteriolar lymphoid sheath (PALS) is a region within the splenic white pulp that is populated mainly by T cells. Lymphoid follicles, some with germinal centres rich in B cells, are seen attached to the PALS.

Chapter 23

THE CELL WHISPERERS

Gothenburg, Sweden, 1980. A 19-year-old woman became unwell with feverishness, joint pain, severe headache, diarrhoea and vomiting. Because of rapid deterioration she was brought to hospital some 36 hours after the onset of her illness. She was confused and in shock: her temperature was 40°C, her blood pressure was 70/55 mm Hg and she had a pulse of 150 beats per minute. There was a foul-smelling discharge from the vagina, and a tampon was removed: she was on day four of her period. Her eyes were inflamed, and a few days later the skin from her palms began peeling. Her tongue started to swell and became red, a condition later called 'strawberry tongue'. She was treated with intravenous fluids, plasma, steroids and antibiotics. After a waxing and waning course, she recovered. Staphylococcus bacteria was cultured from her nose and throat, as well as from the tampon.[1] This is an example of tampon-induced toxic shock syndrome, which was first described in the late 1970s and early 1980s, when several young women died from infections caused by using super-absorbent tampons. Although unknown at the time, the symptoms were the result of an overproduction of small proteins called cytokines. In this chapter, we will see how these important molecules, now understood to be involved in a diverse array of immune functions and pathological processes, were discovered.

Miller and Mitchell's experiments showed that the production of antibody by bone-marrow cells often required the assistance of thymic cells. But how did these cells cooperate? How did T cells help B cells to make antibody? How did they communicate? Although the exact mechanism underlying this cell-to-cell cooperation remains unresolved, the process is known to be mediated by cytokines.

*

In the 1940s, Eli Menkin and Paul Beeson, working in the United States, discovered a substance that caused fever in rabbits exposed to a sterile irritant. Later, this substance, called 'endogenous pyrogen' (EP), was found to be a protein.[2]

In 1971, Igal Gery, working at Yale, isolated thymic lymphocytes from mice immunized with sheep red cells. He then added some phytohaemagglutinin, a plant lectin, into the lymphocytes that had been placed in a dish. Even though phytohaemagglutinin was a 'mitogen', a substance that caused immune-cell division, Gery found that it could not stimulate the thymic lymphocytes to divide unless he added some sheep red cells into the mixture. So did the sheep red cells cause the thymic lymphocytes to divide?

Surprisingly, it wasn't the red cells that caused the cell division by the mitogen. Later experiments found that white blood cells contaminating the sheep red cells added to the culture had caused the cell division. But Gery found that cell-free fluid also caused thymic-lymphocyte division by the mitogen. By carefully separating the different types of white cell and repeating the experiment, Gery found that the substance causing the cell division was produced from macrophages and not from the lymphocyte fraction. When purified, this factor, which he called 'lymphocyte activating factor' (LAF), was found to induce fever: LAF was also an endogenous pyrogen.[3]

Further experiments revealed that LAF was identical to the EP discovered by Beeson and Menken. So this mysterious substance was both a pyrogen *and* caused lymphocytes to divide. And the antigen – the sheep red blood cells – or a mitogen – the phytohaemagglutinin – required the presence of this substance to cause lymphocyte cell division.

In 1979, at the Second International Lymphokine Workshop, held in Ermatingen, Switzerland, LAF and EP were renamed as interleukin 1. Subsequent experiments revealed that interleukin 1 belonged to a family of interleukins (a class of glycoproteins) which had numerous functions in immune responses. Interleukin 1 was found to be capable of activating lymphocytes, enhancing the activity of other cytokines and promoting inflammatory responses. So, for the first time, a substance had been discovered that enabled communication between different immune cells; and one that also caused a rise in body temperature, which, by being inhibitory to most pathogens, aided their destruction.

*

In 1962, Barry Bloom and Boyce Bennett discovered another cytokine, one that inhibited the migration of macrophages. In their first experiment, they placed peritoneal cells from sensitized guinea pigs inside capillary tubes and observed the macrophages migrating and fanning out of the open ends of the thin tubes. This was a normal finding and something that the motile phagocytes can do. However, when specific antigen was added to peritoneal cells, the migration stopped. Why did adding antigen stop the macrophage fanning out?

Further experiments showed that sensitized lymphocytes in the peritoneal cell suspension were responsible for this inhibition of macrophage migration when antigen was added. It appeared that in this case it was the sensitized lymphocytes, which upon re-encountering antigen produced a factor that stopped the fanning out of macrophages. Bloom and Bennett called this substance MIF, or 'macrophage-migration inhibiting factor'. When purified MIF was injected into the skin of naive guinea pigs, it caused reddening and swelling: a typical delayed-type hypersensitivity reaction. This suggested that MIF could also be a mediator of the delayed-type hypersensitivity reaction *in vivo*.[4]

*

In 1962, George Mackaness, an Australian immunologist, found another substance capable of mediating communication between two types of immune cells. Made by T cells, this substance caused the activation of macrophages. Mackaness carried out a series of elegant experiments to elucidate the role of this substance, which involved two pathogens that lived inside cells: *Listeria monocytogenes* and *Mycobacterium tuberculosis*. In an early experiment, he found that mice that had overcome an experimental infection with the latter being immune to a second lethal challenge by the tubercle: this protective immunity was transferable to naive mice by cells taken from the mouse given the primary infection, but not from antibody. So immunity to mycobacteria appeared to involve a cellular response and not one mediated by antibody. So far so good.

But then things were about to get strangely complicated. Buckle up!

Interestingly, if these mice were challenged simultaneously by *Mycobacterium tuberculosis* and *Listeria monocytogenes*, they showed immunity to both types of bacteria. This was unexpected: the mouse infected with mycobacteria had also gained immunity against a second, unrelated bacterium. How on earth was this possible? What was going on?

It was found that this protective immunity only happened if the challenge injection contained *both* the tubercle *and* listeria: if the challenge injection contained only the listeria, protective immunity against listeria was not observed (Fig. 13). This suggested that it was the secondary response generated towards the mycobacterium that was mediating the resistance against the listeria.[5]

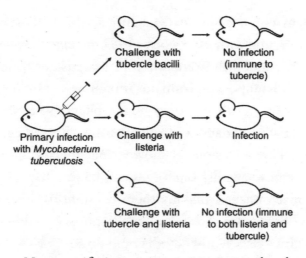

Figure 13. Non-specific immunity against an unrelated pathogen observed following a specific immune response mounted against a pathogen.

This dual protection, conferred against both types of bacteria by a secondary response generated against one type, could also be transferred to naive animals by the cellular fraction but not by antibody. Mackaness wanted to find out which type of cell was responsible. More transfer experiments involving different cell populations revealed that protection was mediated by lymphocytes. Importantly, the lymphocytes that generated the secondary response to the tubercle were responsible for the observed resistance to both the tubercle and listeria.[6] But exactly how did the lymphocytes confer protection?

Both mycobacteria and listeria were intracellular pathogens that had the uncanny ability to survive and thrive inside macrophages: they had evolved mechanisms that prevented them from being destroyed inside the phagocyte. Mackaness found that macrophages isolated from animals who had recovered from a primary tuberculosis infection and were then given a challenge injection of mycobacteria were different from those found in naive animals. They were bigger, looked 'angry' and were able to destroy their unwanted guests.

Similarly, macrophages taken from animals that had recovered from a primary infection with *Mycobacterium tuberculosis*, when exposed to the tubercle in a dish, were found to be resistant to infection by the tubercle if lymphocytes from the animals were also present in the culture. Interestingly these 'angry' macrophages were also resistant to infection by listeria and a third bacterium, *Brucella abortus*. So, collectively, these experiments showed that the specific secondary immune response against one organism had non-specifically activated the macrophages, making them resistant to other, unrelated pathogens.[7] But, as we saw earlier, macrophages could not transfer this immunity – only lymphocytes could do so.

Were the lymphocytes activating the macrophages? And could the lymphocytes, by activating the macrophages, make the animals resistant to infection? Further experiments showed that this was indeed the case: lymphocytes taken from animals previously exposed to a pathogen were able to increase the bacterial-killing activity of macrophages, which were then able to destroy any pathogens living inside them: listeria, *Mycobacterium tuberculosis* and *Brucella abortus*.[8]

Further experiments showed that it was a protein produced by the sensitized lymphocytes that was activating the macrophages. Lymphocytes taken from an animal given a primary infection with tuberculosis and cultured with the tubercle – effectively giving rise to a secondary response in a dish – were able, via this protein, to cause macrophages from a naive animal to change: to become activated and 'angry'.[9]

Taken together, these elegant studies pointed towards a family of small proteins produced from specific T cells that non-specifically activated macrophages, making them able to resist infection even from unrelated pathogens: a clear example of how innate and specific responses were able to work together to destroy intracellular pathogens.

In 1983, this macrophage-activation factor was purified, characterized and named as interferon gamma (IFN-γ). Interferon gamma was a lymphokine, a cytokine made by lymphocytes that activated macrophages during an immune response.

These experiments validated Metchnikoff's indomitable assertion that macrophages were key elements in immunity. His original experiments with the rose thorn and the starfish larvae had demonstrated their main function in innate immunity: phagocytosis. But the innate and specific immune responses were not disparate entities: they joined forces in the battle against pathogens. The discovery of opsonins by Almroth Wright had demonstrated the crucial role of antibody in the enhancement of phagocytosis, and George Mackaness had shown how specific T cells, by producing cytokines, activated macrophages, turning them into lethal phagocytes.

As we saw earlier, T cells can also help B cells make antibody. The mechanism turns out to be a complex one involving several cytokines.[i] Cytokines interleukin 4 and interleukin 21 appear to play a prominent role.

Cytokines can be produced by a myriad of cells such as macrophages, lymphocytes, epithelial cells, fibroblasts and dendritic cells – we will come across this cell in Chapter 29.

i   Cytokines are small proteins that mediate various functions of the cells of the immune system. They are secreted by a wide variety of cell types. Some cytokines activate immune cells, some promote their maturation and others control their movement. By mediating communication between cells, cytokines control almost all immune responses. For example, in some allergic diseases the cytokine IL-4 changes a naive T-cell into a T helper type 2 cell, which, through secretion of IL-4 and IL-13, stimulates B cells to produce IgE antibodies. IL-2 causes T-cell proliferation and, as we saw earlier, interferon gamma secreted by T-cells activates macrophages, allowing them to destroy intracellular pathogens like *Mycobacterium tuberculosis*. The cytokine IL-8 acts as a chemokine, drawing neutrophils to the site of an infection.

Cytokines carry out their actions by binding to cytokine receptors on cells. This binding initiates signals that change the behaviour of the cell. If the cytokine produced its effect by binding to a receptor on the cell that produced it, the effect is called an 'autocrine' effect. If it binds to its cognate receptor on a nearby cell, this is called a 'paracrine' effect, and if the binding involves a distant cell it is called an 'endocrine' effect.

These 'biological response modifiers' have clinical applications. Some cytokines are used to control harmful autoimmune responses: interferon beta, by reducing the number of inflammatory cells, reduces relapses in the autoimmune disease multiple sclerosis. Alternatively, antibodies are used to block certain cytokines that promote undesirable effects: tumour necrosis factor alpha, a cytokine that causes inflammation in ulcerative colitis, is blocked by monoclonal antibodies made against the cytokine. Some cytokines have been used in therapeutic vaccines designed to augment the immune response to tumours. In one recent multi-centre study, lymphocytes infiltrating a tumour from patients with metastatic melanoma were collected, expanded *in vitro* with the cytokine IL-2 and reinfused. Treatment with large numbers of these activated lymphocytes caused tumour regression responses reaching up to 50 per cent, with approximately 10 per cent of the patients showing long-lived or complete remission.

Cytokines produced by the T helper type 1 (Th1) subset, particularly interferon gamma and tumour necrosis factor alpha, activate infected macrophages, making them more capable of destroying the unwanted guest living within it. Cytotoxic T-cells also attack these infected cells. But the canny tubercle has many evasion strategies that enable it to avoid being totally vanquished by the immune assault. Unable to clear the infection, the immune system attempts to lay siege, walling off infected macrophages inside a ball of cells called a 'granuloma'. Inside the granuloma the tubercle waits, contained but dormant – a latent infection waiting to erupt and spread when the host defences have weakened.

The cytokine TNF alpha has been shown to be essential for the maintenance of the granuloma: treatment of rheumatic diseases by anti-TNF-alpha antibody can disrupt the granuloma, causing a reactivation of tuberculosis.

Chapter 24

THE LONG SHADOW OF CHARLES DARWIN

Paris, 1895. Pasteur sat alone in his study, his papers scattered over the large, ornate desk. It was over; his time had come and gone. His career was done. Finally, he had decided to leave the institute. No matter, he thought: it was now in the capable hands of his trusted colleagues Roux, Grancher and Chamberland. They would continue the work; they would make the discoveries.

Then the thought – the infernal thought – occurred. Running his trembling hands through his snow-white hair he began to ponder. He had no choice: the thought could not be resisted.

How did it all happen? How did Jenner's cowpox vaccine work? For that matter, how did his rabies vaccine work? What great miracle had happened within young Joseph Meister when the vaccine was administered? What substances were generated within his body by the vaccine? How did they afford protection against the virulent form of rabies that was injected? How did the boy survive the bite of death? If only he could see inside the boy's body, could gaze at the cells and substances at work.

He had no answer. He didn't know. He would never know. Maybe, he thought, the answers would come to him in a dream. If only he had more time – ten more years, perhaps. But there was no more time: his time had come and gone... It was for another to take up the challenge, to solve the riddle.

*

In the 1950s, Albert Coons, using a fluorescent microscopy technique, confirmed that antibodies were made by the lymphocyte.[1] Subsequent research revealed this cell to be a plasma cell: a mature B lymphocyte capable of churning out copious quantities of antibody. But the mechanisms underlying this process remained obscure.

As we saw in Chapter 11, the first natural-selection theory of antibody production was propounded in 1897 by Paul Ehrlich. In his side-chain theory, Ehrlich suggested that antigen, by binding to a 'side chain' on an antibody-producing cell, selected the specific cell from the many for expansion and antibody secretion. But in the first half of the twentieth century, most immunologists subscribed to a template-instructive theory of antibody formation, one that suggested that antigen acted as a template for antibody production. In 1940, the twice Nobel Prize-winning chemist Linus Pauling proposed that the antibody molecule sitting on the cell surface moulded itself around antigen to give rise to a high-binding antibody that was subsequently secreted by the cell: like a man instructing a tailor to make a perfectly fitting suit especially for him.[2] Landsteiner's work on anti-hapten antibodies supported this theory: there was no other explanation for the pre-existence of cells capable of making antibodies against man-made synthetic compounds that they had never encountered.

But, in the early 1950s, Niels Jerne, a Danish immunologist, challenged the prevailing instructional theory, pointing out that it did not fully explain the memory response, or the increased rate of antibody synthesis seen in the primary response. Antigen had already been introduced, so how could antibody response *increase*? How could a second exposure to antigen lead to a greater and more rapid response?

Jerne proposed a different theory. He suggested, as Ehrlich had done back in 1897, that the information needed to make specific antibodies was already in the genome.[3] Following the discovery of the structure of DNA in 1953, it became clear that information coding for proteins like antibodies resided in DNA, and that this information flowed in one direction: DNA to RNA to protein. Therefore, the information required to make the antibody protein could not reside within antigen or antibody molecules. This information, as Ehrlich originally proposed, must lie within the genome, within DNA.

Jerne suggested that antigen, rather than instructing the formation of antibody, *selects* for production a few antibody types from all antibody molecules already present: like the man in the earlier example choosing a suit that fits him from a selection on a rack. David Talmage, an American immunologist, who was in broad agreement with Jerne's ideas, proposed that antibody-producing cells were being selected and expanded by antigen, and his ideas and those of Jerne were refined and expanded by Frank Macfarlane Burnet, who formulated a comprehensive selective theory of antibody formation, one that was eventually verified by experimentation.

Burnet, born in 1899, was an Australian physician and biologist who worked at the Walter and Eliza Hall Institute in Melbourne. He believed that the antibody molecule was situated on the surface of lymphocytes: Ehrlich had called this a 'side chain'. This molecule acted as a specific receptor for antigen, and, when bound by antigen, caused the cell to divide, producing a clone of daughter cells capable of producing antibody of the same specificity: these daughter cells would also have identical receptors, the antibody molecule, of the same specificity. Antigen thus selects and activates a specific cell from the many, giving rise to an identical clone secreting the same antibody. Burnet's theory, now called the 'clonal selection theory', could also account for the observation of increased antibody production in the primary response: it was all down to the frenetically dividing clone of plasma cells.

To explain memory cells and the secondary response, Burnet proposed the existence of two clones, both with the same specificity, thus capable of making the same antibody. Produced after the initial interaction with antigen, one clone acts immediately to neutralize the pathogen and defeat the infection, while the other, with a longer lifespan, remains dormant to deliver the secondary response if and when the pathogen returns. These cells, the mediators of lifelong immunity, are called memory cells (Fig. 14).[i4] The memory cells were the mediators of the secondary response.

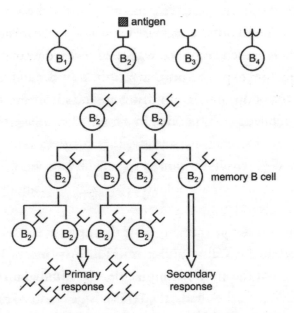

Figure 14. The clonal selection theory of antibody formation.

In 1958, Gustav Nossal, an Australian immunologist who had collaborated with Jacques Miller and the American microbiologist and Nobel Prize-winner Joshua Lederberg, provided the first bit of evidence for Burnet's clonal selection theory by showing that a single B cell was only able to produce one type of antibody.[5]

But Burnet didn't stop there. He went on to propose an explanation for the scarcity of lymphocytes capable of producing antibodies against self-antigens. Such self-reactive lymphocytes, he suggested, would be deleted in foetal life by a rigorous selection mechanism. But autoimmunity did happen, so what was the explanation?

Autoimmunity occurred, Burnet argued, because certain self-anti-gens were somehow hidden from the immune system in foetal life, and the cells capable of reacting to them had thus escaped selection and destruction.[6]

We can look at the nature of the adaptive immune response in the light of Burnet's theory. We know that the first encounter with an antigen leads to a small response and a second to a faster and much larger response. This was intuited from the observations of Jenner, Pasteur and others, and was exploited in their vaccine trials. The same heightened response was seen in Medawar's second set of graft-rejection experiments.

Burnet's ideas on autoimmunity, predicated on the concept of tolerance, were validated by Medawar even if the importance of the lymphocyte in immunity was just becoming apparent at the time when the experiments on acquired tolerance were being conducted. Burnet's theory was finally able to explain how antibody responses were mounted and how the secondary response was generated through the activation of the memory clone. In one fell swoop he had provided mechanisms that attempted to explain the four funda-mental features of specific immunity: *specificity, memory, diversity* and *self–non-self-recognition*. For his contributions to immunology, Frank Macfarlane Burnet was jointly awarded the Nobel Prize in Physiology or Medicine in 1960. The theoretical groundwork of Niels Jerne did not go unrecognized either: he too was awarded a share of the prize in 1985.

Let's take a more in-depth look at how an antibody response is generated towards an antigen – how a typical immune response develops. A similar process occurs in a T-cell response.

As mentioned in Chapter 6, the immune reaction that follows the first encounter with antigen is called the primary response. Naive B lymphocytes, bound by the antigen through its surface antibody receptor, become activated, generating two clones with identical specificity. One clone differentiates into plasma cells, the other into a population of memory cells. The antibody produced by the plasma cells in the primary response is of the immunoglobulin M (IgM) class, and these antibodies typically have low affinity: they don't bind very tightly to antigen. The primary response lasts from a few days to a few weeks, and once completed the individual is said to be sensitized to the antigen. On a second encounter with the same antigen, activation of the dormant memory cells leads to the unleashing of the secondary response. In the first few days more IgM is produced, but soon large quantities of antibodies of other classes appear. These antibodies are produced by memory B cells that have gained the ability to switch from making IgM to making IgG, IgE and IgA antibody. The antibody molecules produced in the secondary response show an increased affinity for antigen (Fig. 15).[ii] This is a consequence of the rapid division of memory B cells. Each time a cell divides, mutational events occur within the genomes of the cells and it is this process, called 'somatic hypermutation', that gives rise to antibodies with greater affinity to antigen. As we saw earlier, antibodies are made from information in the genes (DNA sequences) and mutations in the DNA sequences can alter the structure of the antibody produced. Random mutations can also generate low-affinity antibodies, but memory B cells with high-affinity antibody receptors, by virtue of their increased ability to bind antigen, are selected for expansion. This feature, along with the generation of antibody classes with more efficient pathogen-clearing functions, makes for a more effective response against the antigen. In contrast, the antibodies made during the primary response are found to be closely related showing few somatic hypermutations. It is important to note that this process of somatic hypermutation does not occur in the T-cell response. Diversity of the T-cell repertoire is generated by gene rearrangement, giving rise to different T-cell receptors.

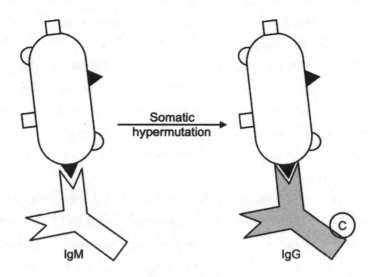

Figure 15. Increased antibody affinity in the secondary response.

So this chapter has partly answered Pasteur's questions. How did it all happen? What is the basis of vaccination? How did the immune system confer protection after vaccination? But we have only scratched the surface. For example, Burnet and the other immunologists tackling these problems didn't know the structure of the antibody molecule. We will look at how this puzzle was solved in the next chapter.

i   Memory B cells show increased expression of MHC class II antigen on their surfaces. As B cells are also antigen-presenting cells, this facilitates antigen uptake and presentation, allowing memory B cells to activate helper T cells at lower doses of antigen.

ii   In addition to undergoing more somatic hypermutations, B cells in the secondary response also produce antibodies derived from VH and VL gene segments not utilized in the primary response. These novel antibodies are thought to be made from small numbers of B cells from the primary response that have been subsequently differentiated and expanded in the secondary response.

Some B cells that have not yet become fully differentiated migrate into the secondary lymphoid organs during the secondary response. There they enter germinal centres, where they encounter antigen and go through additional rounds of cell division. This results in more somatic mutational events and consequently antibodies with better affinity. How exactly did this happen?

In the 1960s, Gustav Nossal and Gordon Ada discovered a germinal centre cell called the 'follicular dendritic cell' (FDC). This cell, which bound and displayed antigen on its surface, played an important role in memory responses. 'It became tempting,' Nossal states, 'to postulate that while B cells were in the germinal centre they revisited the antigen depot repeatedly. Perhaps on each occasion the only B cells which divided were those bearing mutations that raised antibody affinity?'[7]

Three decades later, in the 1990s, Nossal and his colleagues elucidated the process, showing how memory B cells underwent a selection process in the germinal centres of secondary lymphoid organs: this process occurring early in the infection. From the diverse clones of B cells that approached the FDCs, only those capable of binding, via their receptors, to antigen trapped on the surface of the FDC were allowed to progress. These high-affinity B cells were then allowed to return to the bone marrow and produce large amounts of high-affinity antibodies characteristic of the secondary response.

Nossal states: 'As well as producing memory B cells, the germinal centre also produces a special population of antibody-forming B [plasma] cells. These are B cells with raised affinity which are quickly exported from the germinal centre and which migrate to the bone marrow where they live for many months making their important contribution to immunological memory.'[8]

Chapter 25

THE Y-SHAPED BULLET: UNRAVELLING THE STRUCTURE AND DIVERSITY OF THE ANTIBODY

Emil von Behring sat alone in the garden of the Swiss sanatorium he periodically disappeared to when things got difficult – when the dark cloud made its ominous return. Lately he had spent longer periods here amid the lush lawns and the tall cypresses, reflecting, attempting to shake off the depression. He had been awarded the Iron Cross and the first Nobel Prize in Physiology or Medicine. He was a national hero and called the 'Saviour of Children', but his work on serum therapy had exhausted him.

During the First World War soldiers on both sides died an agonizing death from tetanus. The war was fought in fields, and battle wounds were quickly covered in mud, where the toxin-producing bacillus lived. Beginning in 1914, horse serum containing tetanus antitoxin was given to soldiers with wounds contaminated by soil. Antibody therapy drastically cut the death rate; prophylaxis saved thousands of lives. But no one knew what these antibodies looked like. Ehrlich had made drawings of what he thought they looked like: clubs with fishtails. He was off the mark.

*

In 1931, Arne Tiselius, a Swedish biochemist, developed a method that enabled the separation of large molecules like proteins in a mixture. Known as 'electrophoresis', it involved passing an electric current through a gel-like matrix containing wells, into which mixtures of biological molecules were added. Once the current was switched on, different molecules, due to their different sizes and electrochemical properties, moved at different speeds through the gel, enabling the separation of the mixture into different fractions, each containing a group of molecules with an equivalent size and charge.

Teaming up with the American biochemist Elvin Kabat, Tiselius used the method to separate the proteins in serum into different fractions, and demonstrated, in 1935, that antibody resided in the gamma globulin fraction.[1] The crucial experiment involved electrophoresing immune serum from mice immunized with the carbohydrate coat (polysaccharide) of the pneumococcus bacterium. If the immune sera were first absorbed by pneumococcal polysaccharide, and the serum without antibody that had remained stuck to the polysaccharide was electrophoresed, the gamma globulin fraction was reduced. This suggested that the gamma globulin fraction had bound antigen: antibodies against the polysaccharide coat of pneumococci must therefore reside within that fraction.

*

Rodney Porter, an English biochemist, took these studies further. A quick glance at Figure 16 may help you follow these crucial experiments. By treating the gamma globulin fraction with the protein-splitting enzyme papain, Porter was able to split the fraction into three smaller fragments: two identical fragments that combined with antigen and a third that could not. The antigen-binding fragments were called 'Fab fragments'; the third fragment which crystallized was called the 'Fc fragment'. He then used a second enzyme, pepsin, to cut the Fc fragment, obtaining a larger divalent antigen-binding fragment (with two antigen-binding regions) which he called Fab2, and a smaller fragment which was the remaining bit of the Fc fragment.[2]

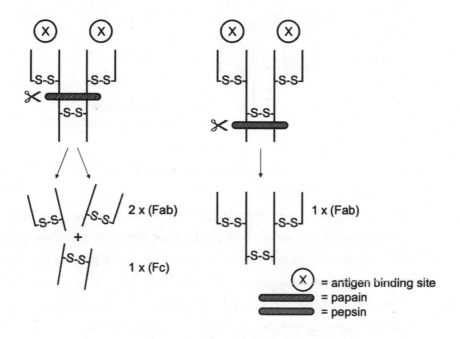

Figure 16. Elucidation of the structure of antibody.

At the end of these enzyme-digestion experiments, Porter was able to propose the composition of the main type of antibody molecule – now called 'immunoglobulin G' or 'IgG' – found in the gamma globulin fraction. Proteins are made up of building blocks of amino acids strung along in chains, like beads on a string. He claimed that IgG was made up of two different chains, joined together by disulphide bonds. Gerald Edelman, an American biologist, further characterized the chains using reducing agents. His experiments revealed that the antibody molecule was composed of two types of chain: a 'light chain' and a 'heavy chain'.[3]

These experiments allowed Porter and Edelman to propose a model for the antibody molecule. IgG was a large molecule composed of four chains: two heavy and two light, joined in the shape of the letter Y (Fig. 17). Later electron microscopy showed that IgG was indeed a Y-shaped molecule.[4] Finally, Behring's two-pronged projectile antitoxin antibody could be visualized.

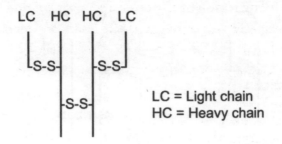

Figure 17. The Structure of IgG.

Porter and Edelman wanted to analyse the structure of the antibody molecule in more detail, but this required large quantities of *identical* antibody. Fortunately, they found a human disease in which patients produced identical antibody molecules in abundance: multiple myeloma, a type of blood cancer. In this disease, a single B cell became cancerous and began multiplying uncontrollably to give rise to a clone of identical cells, all producing a single type of antibody called a myeloma protein. Some myeloma proteins, dimers of light chains, eponymously called 'Bence Jones proteins', are also frequently found in the urine of these patients.

When rabbits were immunized with a variety of Bence Jones light chains, two distinct types of anti-light-chain antibody were produced. This meant that there were two different types of light chain, and these were characterized and subsequently named 'kappa' and 'lambda'. Similar experiments involving analysis of antibodies raised against a range of myeloma and normal antibodies revealed the existence of five heavy-chain classes: gamma, mu, epsilon, alpha and delta.[5]

Determination of the amino-acid sequences of Bence Jones proteins and myeloma heavy chains revealed the existence of two distinct regions in both light and heavy chains: a variable and a constant region.[6] Variable regions have lots of different types of amino acid in the various myeloma chains, whereas constant regions have much less variability in their amino-acid composition. As expected, the variable regions of the light and heavy chains jointly formed the antigen-combining region of the molecule and explained the diversity of the antibody response. In 1969, Edelman determined the amino-acid sequence of an entire antibody molecule.[7]

Amino-acid sequencing of many antibody molecules by Wu and Kabat in 1970 revealed three to four regions with even more variability, termed 'hypervariable regions' within the variable regions; these were subsequently shown to be the specific regions in the antibody molecule that bound antigen.[8]

In 1950, Linus Pauling had suggested that the exquisite specificity of the antibody molecule was due to the existence of a three-dimensional structural complementarity between antigen and antibody,[9] and the experiments carried out by Porter, Edelman, Wu, Kabat and others had shown how the structure of the antibody molecule made this possible. The discovery of hypervariable regions also explained the presence of large numbers of antibody specificities binding to an indefinite number of antigens. Finally, the structural basis of the specificity of the antibody response had been experimentally demonstrated. For the discovery of the structure of the antibody, the 1972 Nobel Prize in Physiology or Medicine was awarded to Porter and Edelman.

There are five classes of antibody. IgM, the large pentameric antibody, is the class that appears during the primary response. Interestingly, it is also the first antibody that is made by the foetus. A mechanism called 'class-switching' then produces IgA, IgG, IgD and IgE, and these antibody classes are seen in the secondary response. They have different biological functions; for example, IgG can cross the placenta, affording protection to the foetus.

We can now summarize the functions of antibodies. Some neutralize toxins, like Behring's antitoxins against diphtheria and tetanus, the antibodies that saved countless lives. Other antibodies coat bacteria and make them more readily phagocytosed: Almroth Wright's opsonin. These bacteria-binding antibodies can also activate the complement cascade which causes bacteriolysis and the generation of the inflammatory response. Some antibodies can physically block antigens from binding to host tissues. For example, secretory IgA antibodies can block pathogens from attaching to the epithelium of the gastrointestinal tract.

*

It was becoming clear that there were millions of B and T cells, each with a unique receptor, giving rise to an antibody and T-cell repertoire capable of binding with an unimaginable number of different antigens. This led to the question: how was this tremendous diversity achieved?

Early theories on the generation of diversity focused on the role of somatic mutation of a few antibody genes. These gene mutations, the theory suggested, created the vast numbers of antibody molecules. But this did not fit with several observations. The vast repertoire was present quite early in development: insufficient time for enough somatic mutations to occur to generate the observable level of diversity. Also, foetuses and newborns appeared to develop antibody responses to different antigens at precise stages of their development, and in a precise order. Thus, the mechanism responsible for antibody diversity was unlikely to involve a random somatic mutational process.

In 1965, J. Claude Bennett and William Dreyer from the California Institute of Technology (Caltech) came up with an alternative theory.[10] The variable and constant regions of the antibody molecule, they suggested, were encoded by two separate genes. These two genes, called V and C genes, combined during the development of lymphocytes to produce the antibody molecule. A decade later, a Japanese immunologist, Susumu Tonegawa, showed exactly how this happened.

*

Tonegawa was born in 1939 in Nagoya, Japan, just five days after the outbreak of the Second World War. In 1959, having failed at his first attempt, he managed to secure a place at Kyoto University. After graduating with a degree in chemistry, Tonegawa decided on a research career in the new field of molecular biology. Needing to get a PhD first, he began looking for a suitable university. But this was a tumultuous period for post-war Japan and Tonegawa was advised to leave for the United States to pursue his doctoral studies. He managed to secure a place at the University of California, San Diego, and in 1968 received his doctorate in molecular biology. He then began post-doctoral work at Renato Dulbecco's laboratory, which he described as 'the best laboratory in the field of molecular biology'.[11]

After gaining a solid grounding in molecular biology, Tonegawa decided to switch fields and enter the growing field of immunology, and in the winter of 1971 left America to work at the Basel Institute of Immunology under the direction of Niels Jerne. There, he writes, 'I was introduced to the great debate on the genetic origins diversity of antibody diversity.'[12]

Tonegawa was on a mission. He was going to solve the riddle of how the immune system achieved such a remarkable diversity: how so many different antibody molecules, able to bind to a seemingly infinite number of different antigens, could be produced. He felt that he could use the recently developed techniques of molecular biology – restriction enzymes and recombinant DNA – to solve the problem of how diversity was generated. He did.

Tonegawa left Basel in 1981, having resolved the great mystery: the mechanism by which antibody diversity occurred. He showed that there were two separate gene loci coding for variable and constant regions of the light chain. However, these gene loci were made up of smaller gene segments. This was suggested by the fact that the variable part of the light chain had more amino acids than could be encoded by just one variable (V) gene segment, suggesting that another gene segment was involved in the coding for these extra amino acids. This was the J or 'joining segment'. Thus, the V and J gene segments joined together to encode the variable part of the light chain. It was subsequently shown that the heavy-chain variable region was also put together by the recombination of several gene segments, called V, D and J segments. During development, these variable gene segments encoding both light and heavy chains combined randomly with constant gene segments to give rise to the vast repertoire of antibody molecules (Fig. 18). So it was all down to a reshuffling and recombination. The gene segments can be thought of as a pack of cards. There are a fixed number of cards in a pack and the pack can be reshuffled to produce a hand of cards. The particular hand given out is a particular antibody.

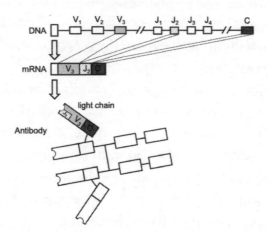

Figure 18. Generation of antibody diversity (adapted from Kenneth Murphy and Casey Weaver, Janeway's Immunobiology (9th edn, New York: Garland Science)).

To make things even more complicated, the variable gene segments themselves were also found to be highly variable. Tonegawa showed that there were 50 distinct types of variable kappa-gene segment and two distinct types of variable lambda segments in mice, and by combining these different V-region gene segments with a single C region it was possible to generate a wide range of light chains. The same was true for the heavy chain.[13]

Interestingly, the gene segments coding for light and heavy chains were found to reside on different chromosomes. In the human, the lambda light-chain gene segments are found on chromosome 22, the kappa light-chain gene segments on chromosome 2 and the heavy-chain gene segments on chromosome 14. Similarly, in the mouse the kappa, lambda and heavy-chain gene segments are all found on different chromosomes.

For working out how antibody molecules were put together, how diversity was achieved, Susumu Tonegawa was awarded the 1987 Nobel Prize in Physiology or Medicine.

Tonegawa showed us how antibody diversity was generated. But what about the cellular response? Like the myriad antibody molecules, there are millions of different T cells, each armed with a receptor capable of binding specific antigen. Thus, diversity is also seen in the cell-mediated arm of immunity. This meant that there was a T cell ready to engage any virus that was around, even one that the human immune system had not yet encountered, one waiting in some dark, bat-infested cave in the depths of a Chinese forest.

How is this possible? As with the antibody, it is down to the T-cell receptor. This receptor, projecting out of the cell membrane, is made up of two polypeptide chains: an alpha and a beta chain, which, like those of the antibody molecule, have constant and variable regions. And, as in the case of the antibody, the variable and constant regions are encoded by multiple gene segments which recombine to generate a very large number of different T-cell receptors, all capable of binding to different antigenic peptides. As each T cell has only one unique receptor, this process leads to the production of a very large number of T cells capable of binding to a limitless array of antigens.[i]

Gene-segment recombination, which precedes production of antibody or T-cell receptors, is carried out by recombination-activating enzymes, encoded by recombination-activation (RAG) genes. The RAG enzymes do this by cutting the DNA of V, D and J gene segments, generating gene fragments that recombine randomly to give the final recombined VDJ gene complex. It is this cutting, shuffling and rejoining that gives rise to millions of different VDJ gene complexes, and consequently millions of different antibody and T-cell receptors. As gene recombination is required for the generation of diversity, expression of RAG genes is restricted to developing lymphocytes; in fact, expression of RAG genes appear to be essential to the generation of mature T and B cells.

It is important to note that somatic mutations in these gene segments play a role in the secondary response. As the B cells divide to form the clones of memory cells that give rise to the secondary responses, mutational events occur giving rise to antibodies that bind more tightly to antigen. Thus, more effective antibodies are found in the secondary response. Mutations can also give rise to antibodies that bind more weakly; but the better-fitting antibodies, sitting atop memory B cells for example, by selectively binding to antigen get selected for expansion.

i   T-cells acquire a functional alpha-beta ($\alpha\beta$) T-cell receptor (TcR) complex in the thymus. Like the antibody genes, the TcR genes show intense genetic rearrangement that creates millions of different T cells with different TcRs, thus generating diversity. Most of the T cells express alpha-beta TcRs ($\alpha\beta$ T cells), but some T cells found in epithelial tissues (like in the gastrointestinal tract) express gamma-delta TcRs ($\gamma\delta$ T cells).

Chapter 26

THE IMMORTAL CELL

Basel, Switzerland, 1973. Georges Köhler, a German immunologist, approached the biochemist who had just delivered a spellbinding presentation at the Basel Institute of Immunology. Köhler, a quiet, softly spoken man, the archetypal introverted bench scientist, had been captivated by the speaker from Cambridge. Soon the two men would join forces to develop a technique that would lead to revolutionary new weapons against cancer and autoimmune disease, one that had a dizzying array of applications, from medical diagnostics to agriculture. The man Köhler approached was César Milstein.

*

In 1958, Milstein, the son of Jewish immigrants who had migrated from Ukraine to Argentina, left for England on a British Council Fellowship. After carrying out research in enzyme biochemistry, he gained a doctorate – in just two years – from the University of Cambridge and began working at the Medical Research Council (MRC) laboratory in the department of biochemistry at Cambridge.

After introducing himself, Köhler asked Milstein if he could join his team, and in April 1974 Georges Köhler arrived in Cambridge. Their collaboration revolutionized immunology, and led to both men, along with Niels Jerne, being awarded the Nobel Prize for Medicine in 1984. The prize was awarded to Köhler and Milstein for the development of a single experimental technique.

In his Nobel Lecture in 1984, Milstein explained the background to his work:

When an animal is injected with an immunogen [antigen] the animal responds by producing an enormous diversity of antibody structures directed against different antigens, different determinants of a single antigen, and even different antibody structures directed against the same determinant. Once these are produced they are released into the circulation and it is next to impossible to separate all the individual components present in the serum. But each antibody is made by individual cells. The immortalization of specific antibody-producing cells by somatic cell fusion followed by cloning of the appropriate hybrid derivative allows permanent production of each of the antibodies in separate culture vessels. The cells can be injected into animals to develop myeloma-like tumours. The serum of the tumour-bearing animals contains large amounts of mono-clonal antibody.[1]

So what were Milstein and Köhler after? They wanted to develop a method that allowed for the isolation of a single clone of identical plasma cells capable of producing an antibody of a single specificity.

Before Köhler and Milstein's discovery, a single plasma cell couldn't be isolated and expanded by cell division to produce a clone of identical cells. This was because it was pretty much impossible to grow plasma cells in culture. Tumour cells, on the other hand, could be grown in culture as they had the property of 'immortality'. Malignant plasma cells, called myelomas, that could be grown indefinitely in tissue culture had been discovered in the 1960s. Could these cells be used to pull out a single plasma cell capable of dividing and producing an identical clone, making a single antibody?

Köhler and Milstein found a way to grow plasma cells in culture. They had to be 'immortalized'. They did this by 'fusing' plasma cells with myeloma cells, producing hybrid cells called 'hybridomas' that contained the genetic information of both types of cell. These hybridoma cells had the infinite proliferative capability (immortality) of the myeloma cell *and* the ability to make specific antibody, which they got from the plasma cell. Individual hybridoma cells were then isolated – by working out the volume that contained one cell – and grown separately to produce a clone of cells capable of generating an infinite amount of antibody with unique specificity. A single antibody-producing plasma cell had been immortalized, fished out of the mix and expanded.[2]

Köhler and Milstein's original method was elegant in its simplicity – a method even I was able to carry out many years ago. It first involved immunizing a mouse with a strong antigen: sheep red blood cells. After a few months they sacrificed the animal and removed the spleen, which was teased apart to obtain a cell preparation, one that contained plasma cells making antibodies against sheep-red-cell antigens. These splenic cells were then fused with a mouse-myeloma cell line, one that didn't produce antibody. Fusion was achieved by slowly adding inactivated Sendai virus into a spleen and myeloma cell mixture. The virus dissolved the cell membranes of the two cell types, causing them to fuse together to produce a hybrid cell. But in the mixture, there were unfused spleen cells and 'myeloma–myeloma hybrid cells'. The spleen cells would die within a week, but what about the unwanted myeloma cells? Köhler grew the fused cells in a special medium containing hypoxanthine-aminopterin and thymidine – called HAT medium for short – that selectively killed off the unfused myeloma cells and any myeloma–myeloma hybridomas, leaving behind the spleen–myeloma hybrid cells or hybridomas (Fig. 19).

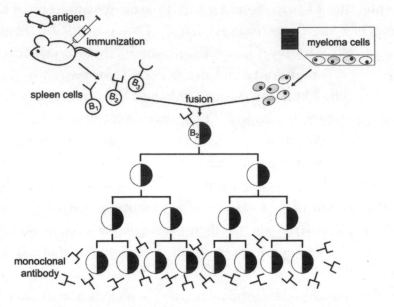

Figure 19. The production of monoclonal antibody.

After carrying out the fusion, the cells were counted and the volume containing just one cell was estimated. This volume was then carefully pipetted into a 96-well microtiter plate, each well potentially containing just one hybridoma cell. But how could they tell if the hybridoma clones growing like clusters of pearls in the wells were making antibody? Luckily there was a new plaque cell assay that allowed such cells to be identified. This assay involved plucking out the hybridoma clones from the wells and spreading them on an agar plate, which was then overlaid with a suspension of sheep red cells. Any antibody-producing hybridoma clones would secrete anti-sheep-red-cell antibody that would bind the sheep red cells, and, when complement was added, lyse them. This would show up as a greenish ring around the hybridoma cell.

Around 5 p.m. on the day of the final phase of the experiment, Georges Köhler added the hybridoma cells onto the agar plates, overlaid them with red blood cells and went home, as he knew that the process would take quite a few hours to develop. But, restless and impatient, he returned to the lab a couple of hours later, along with his wife, to look at the plates. Inspecting them, Georges Köhler saw green halos around some hybrid cell colonies: ringed islands set in a sea of blood-red agar. There they were, the hybridoma cells, producing lots of specific antibody. They had finally found a way to pluck them out and make them immortal. He was so excited that – apparently – he began shouting and kissing his somewhat bemused wife![3]

So if a mouse was immunized with a mixture of antigens, it was possible, by fusing its spleen cells with myeloma cells, to isolate a single plasma cell producing a unique antibody against one antigen out of the many in the mixture. The age of the monoclonal antibody had arrived. This method also provided evidence for the clonal selection theories of Niels Jerne and Frank Macfarlane Burnet. A single plasma cell (a mature B cell) makes one antibody with a unique specificity.

Since their discovery, monoclonal antibodies have been extensively used as diagnostic and therapeutic agents. However, mouse monoclonal antibodies are not ideal therapeutic agents: when introduced into a human the mouse antibody, being a foreign protein, provokes an immune response. Monoclonal antibodies of human origin would be an ideal alternative, but these are notoriously difficult to make using existing technology. In order to get around the problem, human–mouse chimeric antibodies were developed. These chimeric antibodies, part mouse and part human, appear to be better tolerated in patients.[i]

It is interesting to see how these chimeric monoclonal antibodies are produced. Warning: what follows is a bit complicated, so perhaps a coffee break might be a good idea before reading on.

As before, the first step involves immunizing a mouse with the desired antigen, for example the cytokine, tumour necrosis factor (TNF) alpha. The mouse splenic plasma cells will dutifully churn out TNF-alpha antibody.

In the next step, a hybrid gene is constructed from mouse and human DNA sequences. To do this, the DNA sequence coding for the variable region of the anti-TNF-alpha antibody is isolated from plasma cells from the immunized mouse, and this gene is fused with the DNA sequence coding for a human constant chain obtained from human cells. The hybrid gene thus produced can then be put into a bacterium and expanded. The bacterium, basically an antibody factory, would secrete substantial amounts of chimeric human–mouse monoclonal antibody, which can be purified and made into a drug. As TNF alpha is a pro-inflammatory cytokine, monoclonal antibodies against TNF alpha have been used as an anti-inflammatory agent in several inflammatory diseases including rheumatoid arthritis and inflammatory bowel disease. But this novel drug is not without side effects, which can include an increased susceptibility to infection and a reduction in immune cell numbers.

Monoclonal antibodies can also be used to treat several cancers, and this approach has been successful in inducing remission in several intractable forms of the disease. New anti-cancer monoclonal antibodies are being produced regularly, with apparent breakthroughs being reported almost monthly. Reminiscent of Ehrlich's Salvarsan, some of these magic bullets have shrunk stage-four tumours, pushed patients into long-term remission and given hope to those in whom all other treatments had failed: patients previously deemed terminal.

Some of these monoclonal antibodies target tumour-specific antigens, antigens only found on tumour cells, and therefore, in theory at least, they should not attack normal cells. These therapeutic antibodies can be used unmodified or they can be coupled with an agent to increase their killing power. For instance, toxins, chemical agents and radioactive particles can be attached to a monoclonal antibody specific for a tumour-specific antigen. Injecting the patient with such a monoclonal antibody would deliver the antibody to the tumour cell, which would be destroyed by the attached radioisotopes or toxins. Due to their exquisite specificity, these conjugated antibodies would leave normal tissue unharmed.

*

In 1981, Ronald Levy, an oncologist from Stanford, became interested in using monoclonal antibodies against cancer. Coming across Milstein and Köhler's discovery he wrote:

> It totally revolutionized everything I was thinking about being able to do. I adopted their technology immediately, and we made antibodies against human cancer cells. These were initially used for analysis, diagnosis, and characterization, but I got the idea that it might be good to use them as a way of treating cancer.[4]

Levy and his colleagues decided to use the innovative technology to treat a 64-year-old man with terminal B-cell lymphoma, a cancer that had metastasized to the liver, spleen, bone marrow and peripheral blood. The prognosis was grim. But, because this was a lymphoma that had originated from a single cell that had produced a clone of identical B cells, all the cancer cells had an identical membrane-bound antibody on their surfaces – the perfect target for a monoclonal antibody. Levy and his associates produced a mouse monoclonal antibody against this unique B-cell receptor, which, when injected into the patient, sought out and bound to the B lymphoma cells. Like a heat-seeking missile speeding towards its target, the antibodies homed in on the malignant B cells and nothing else. The binding of the monoclonal antibody to its target on the cancer B cell caused activation of complement, which resulted in lysis of the tumour cell: a perfect kill with no collateral damage. After four injections with this anti-B-cell monoclonal antibody, the patient's tumours began to shrink, and he went into remission. It was nothing short of a miracle.[5]

But this approach has practical limitations, as each monoclonal antibody must be tailor-made individually because every patient will have their own unique B-cell lymphoma. Not only would this approach be prohibitively expensive, it would also be time-consuming. But a less custom-made approach can be used, and this is because all activated B cells, whether normal or cancerous, bear certain B-cell-specific surface antigens. One such marker, CD20, is found on all B cells. Rituximab, a monoclonal antibody generated against CD20, has been used to treat non-Hodgkin lymphoma. Rituximab has also been used to eliminate excessive or unwanted B cells from leukaemia and autoimmune disease such as rheumatoid arthritis. Of course, this approach will target all B cells, but once the cancerous cells are destroyed and the drug is stopped, new healthy B cells, formed from the bone marrow, will replenish the stocks.

A different approach involves targeting normal antigens that have been expressed inappropriately in cancer cells. A variety of tumours express higher than normal levels of growth factors or their receptors on their surfaces, and these antigens have become targets for monoclonal antibodies.

For example, an increased expression of human epidermal-growth-factor-like receptor 2 (HER2), encoded by the *neu* gene, has been found in the tumour cells of nearly a third of women with metastatic breast cancer. By contrast, HER2 is expressed only in trace amounts in normal cells. Exploiting this difference in receptor density, a monoclonal antibody against HER2 has been used to treat HER2-expressing breast cancers.

Another, more recent approach involves blocking inhibitory receptors found on T cells. These 'brakes' on the T cell are there to control unwanted T-cell activation, preventing them from turning rogue and attacking self-tissue. As such they are permanently switched on in regulatory T cells, and mutations in the genes encoding these molecular brakes have been shown to lead to autoimmunity.

But sometimes the brakes must be released to allow the T cells to attack a cell we need to eliminate, a cell that has become cancerous. A recent report showed dramatic results when monoclonal antibodies targeting two of the brakes (PDI and CTLA4) were used in the treatment of a particularly lethal type of skin cancer: malignant melanoma. Two monoclonal antibodies, ipilimumab and nivolumab, were used in an international trial involving 945 patients.

The study showed that the treatment stopped the spread of the cancers in 58 per cent of cases. This is an incredible result considering that these were stage-four tumours and the patients were terminal.

The chief of medical oncology at Yale Cancer Center in the United States, Professor Roy Herbst, was quoted in *The Guardian* as saying that 'the treatment, which uses the body's immune system to attack cancerous cells, could potentially replace chemotherapy as the standard cancer treatment within five years. I think we are seeing a paradigm shift in the way oncology is being treated... The potential for long-term survival, effective cure, is definitely there.'

The report also quoted Dr James Larkin, a consultant at the Royal Marsden hospital and one of the lead investigators in the UK:

By giving these drugs... you are effectively taking two brakes off the immune system rather than one, so the immune system is able to recognise tumours it wasn't previously recognising and react to that and destroy them. For immunotherapies, we've never seen tumour shrinkage rates over 50% so that's very significant to see. This is a treatment modality that I think is going to have a big future for the treatment of cancer.

The article gives details of one patient who has benefited from the treatment: a former college teacher, Vicky Brown, 61, was told in 2013 she had only months to live after her skin cancer spread to her breast and lungs.

She took part in clinical trials at the Royal Marsden hospital last August and within weeks the tumour was eradicated. When it later returned, it was removed with immunotherapy. It has come back a third time but doctors plan to use the same method.

'I started drug therapy in August 2014, and although I experienced quite severe side effects these were able to be treated so I could stay on the trial... One of the lumps disappeared after just a couple of weeks which was remarkable. I was delighted to be given the chance to join this trial, not only for me, but also for all the other melanoma patients who could benefit in the future.'[6]

i   Chimeric antibody contains the constant region of human heavy chain and the variable region of a mouse heavy and light chain.

Chapter 27

THE GENES OF IMMUNITY

Pumwani, Nairobi, 1985. Layla sat on a stool outside her small, tin-roofed shack and waited for the men. Her mother had collected her two young children that morning and her parting words were still ringing in her ears. 'You will get sick,' she told her. 'You will get sick and you will die. Who's going to look after the boys? I'm old and haven't got much time left.' Layla said nothing and gazed at the tall buildings of Nairobi in the distance. She had heard this before. Since the new disease had swept through Pumwani many of her friends had got sick and died. But every morning she sat on her stool and waited. She was popular and she had plenty of regulars. Most of them were nice to her, but refused to wear a condom, and Layla knew the risks. The disease was spreading everywhere, all over the world. People were dying; they didn't have any medicines for it.

But Layla kept working: most days, for years and years. Her mother died in 1990, and Layla left her sons with a friend and carried on working. She never contracted HIV. Almost all the other girls she knew in Pumwani got sick and died. But she survived.

One day some doctors came to the village. They took blood samples from all the working girls, every six months for years. She asked one of the doctors why she never got sick. The doctor shrugged. 'We don't know,' he said. 'Some girls never catch the virus. You're one of the lucky ones.' He asked her how long she has been working. 'Nearly ten years,' Layla replied. The doctor told her that it was the girls who worked the longest and saw the greatest number of clients who seemed not to catch the virus. 'How many clients do you see in a day?' he asked. 'Around four, sometimes a few more,' Layla muttered. After that day Layla didn't see that doctor again.

She wandered off, thinking about her friends who had died. Most of them had fallen ill quickly, becoming too sick to work. After about four years they got really sick and died.

The Pumwani cohort was followed up and tested for decades.[1] While most contracted HIV, some didn't. They were persistently negative for the virus and didn't make antibodies. They were resistant to infection. It appears that this protection is down to a set of genes regulating immunity. How were these genes discovered?

*

Paris, 1952. The wounds of war were beginning to heal, and life was returning to normal in the once-occupied city. Jean Dausset, a French immunologist, was preparing to carry out an experiment. But resources were limited, and Dausset worked alone in a small and sparse laboratory with just a bench, a microscope and an old fridge. But the simple experiment would lead to the elucidation of one of the biggest questions in immunology: the genetic basis of immune responses.

Dausset was born in Toulouse, France, in 1916, and, at the age of 11, moved with his family to settle permanently in Paris. His father, Henri Dausset, one of the leading rheumatologists in France, encouraged Jean to pursue a career in medicine. But the advent of the Second World War interrupted his medical studies, and young Dausset was mobilized in 1939. When Paris was occupied by German forces in 1940, he left France and joined the Allied forces in North Africa. There, during the Tunisian campaign, he performed his first blood transfusions, and in Algiers he carried out his first laboratory experiments on blood platelets. In 1944, he returned to a liberated Paris and began working for the Regional Blood Transfusion Centre at Hôpital Saint-Antoine, where he was responsible for the collection of blood for transfusion.

His chosen clinical specialisms were haematology and paediatrics, but Dausset was drawn to laboratory research. He began his research career in the early 1950s, in the new discipline of immuno-haematology, and in 1958 was appointed head of the Immuno-Haematology Laboratory and assistant professor of haematology at the Faculty of Medicine in Paris. In 1953, he became professor of haematology, and, in 1963, he was put in charge of the Immunology Department at Hôpital Saint-Louis. In 1977, the Collège de France conferred on him the Chair of Experimental Medicine.

On that chilly day in 1952, Jean Dausset, driven by curiosity, carried out a rather odd experiment. Interested in the diverse types of antibody found in patients who had undergone multiple blood transfusions, he mixed white blood cells from one patient with a serum sample prepared from another patient, one who had received multiple blood transfusions. The white blood cells agglutinated: antibodies in the sera were binding to antigens on the white blood cells, causing them to clump together. So it appeared that antibodies generated because of multiple blood transfusions were capable of binding to white blood cells belonging to another individual. Intrigued, Dausset repeated the experiment, mixing the white cells with sera from several recipients who had received multiple transfusions from the same donor. A pattern emerged.[2] Realizing that he was onto something, Dausset continued, using antibodies found in the blood of transfusion patients as probes to define the antigens on white blood cells against which they reacted. The experiments eventually led to the characterization of a set of antigens called 'human leucocyte antigens' (HLAs) found on white blood cells, and, as was discovered later, on many other cells. These antigens, found to be responsible for graft rejection, were encoded by a gene complex that was eventually called the 'major histocompatibility complex' or MHC.

What were these strange antigens? What were their biological functions? Soon it emerged that there was a practical application for Dausset's findings. By carefully observing the pattern of agglutination, it was possible to determine the set of human leucocyte antigens found on cells from a particular person: this was the person's HLA or tissue type. And HLA tissue typing allowed close matching of donor and recipient, a method that eventually revolutionized organ transplantation. If you recall, Donnall Thomas had initially used skin grafting to see whether a bone-marrow transplant would work between genetically dissimilar individuals. He had closely followed Dausset's work and had eventually adopted more precise tissue-matching methods.

Dausset's work had been preceded by studies carried out in mice. Peter Gorer, an English pathologist working at Jackson, Maine, in the 1930s, had found that tumours could be transplanted, without rejection, between unrelated mice that shared a set of genes, which he called 'antigen 2'. But they would reject grafts from mice that had a different antigen-2 gene complex.[3]

In the late 1940s, George Snell, an American geneticist and transplantation immunologist and a colleague of Peter Medawar, also carried out a series of mouse-breeding experiments, resulting in pretty much the same finding as Gorer: transplantation rejection was controlled by a gene locus, which Snell called the 'H locus'.[4] In fact, Avrion Mitchison, Medawar's research student, had used some of Snell's inbred mice in his early experiments, some of which we have already come across.

Gorer and Snell had stumbled upon the same locus. Antigen 2 and the H locus were identical. Eventually they agreed to call this locus the 'H2 complex'. Thus, transplant rejection in mice was controlled by a single gene locus: the H2 complex.

But what about humans? Was there, in humans, a similar gene complex that controlled transplant rejection? This question, as we have already seen, was answered by the experiments of Jean

Dausset. The HLA or MHC complex was the human equivalent of the mouse H2.

These studies and Peter Medawar's exhaustive rabbit skin-graft experiments had shown that the immune rejection of allografts was under genetic control. But these H2 or HLA antigens were not sitting on lymphocytes to exasperate transplantation immunologists. Could they have other immunological functions?

The answer to this important question came from the work of the Venezuelan immunologist Baruj Benacerraf, who discovered an important clue as to why some people were more likely to get infections, certain cancers and autoimmune diseases, while others were more resistant to such diseases. He found that it was down to the MHC, which, by controlling immune responses against certain antigens, determined susceptibility and resistance.

*

Benacerraf, of Spanish Jewish ancestry, was born in Caracas in 1920. His father was a textile merchant and importer. The family moved to New York in 1940 and Baruj attended Columbia University, graduating with a bachelor's degree in science. His dream, however, was to study medicine, but getting into medical school was no easy task. In his own words, 'I did not realize, however, that admission to Medical School was a formidable undertaking for someone with my ethnic and foreign background in the United States of 1942. Despite an excellent academic record at Columbia, I was refused admission by the numerous medical schools I applied to.'[5]

He persevered, and was finally admitted to the Medical College of Virginia in Richmond, graduating in 1944. But the world was at war, and Baruj joined the army. He was posted to Europe, where he worked as a community doctor until his discharge in 1947. Then, 'motivated by intellectual curiosity', he decided upon a career in medical research.[6]

Benacerraf began his pioneering work in the new field of immunogenetics at New York University. He was interested in examining the genetic basis of immune responses. It is important to bear in mind that, at this time, the MHC was thought to be concerned with graft rejection and self–non-self-recognition and was not considered particularly relevant to immunity against infectious diseases.

While it was known that some individuals were weak responders to certain antigens, the basis for this observation was unknown. Was it simply a matter of genetics? And, if that were the case, which genes controlled immune responses to these antigens? These were the questions that Benacerraf tried to answer. But as antigens were complex, eliciting varied antibody responses, it was difficult to analyse specific immune responsiveness to complex antigens. But this problem could be circumvented by examining antibody responses to simple synthetic chemicals: Landsteiner's haptens.

In one of his early experiments, Benacerraf found that 40 per cent of the guinea pigs injected with a synthetic hapten, poly-L-lysine, failed to mount antibody and delayed-type-hypersensitivity responses against this antigen.[7] More experiments followed, and the response of outbred guinea pigs to the synthetic antigen was found to be under the control of a single autosomal dominant gene. Benacerraf found that those animals having the gene were responders, and those not having the gene were non-responders. He called such genes immune 'response genes' or 'Ir genes'.

Benacerraf injected two inbred strains of guinea pig, strain 2 and strain 13, with poly-L-lysine. The results were unambiguous: group-2 strain gave a good antibody response against the hapten, but group-13 strain guinea pigs did not. First-generation animals (that is, a 2×13 cross) gave a vigorous response. Further studies found that this observation extended to other species as well.[8]

The availability of inbred animals allowed the rapid mapping of immune response genes. Responsiveness of inbred mice to several proteins was found to be predicated by their unique H2 genotype. Subsequent studies found that Ir genes were found within the H2 complex. This brought the work of Dausset, Gorer, Snell and Benacerraf together, and led to the characterization of a gene locus responsible for immune responses against foreign antigen *and* the rejection of grafts.

The major histocompatibility complex (MHC) is a large cluster of genes. In humans, this complex is found on the short arm of chromosome 6. We now know that the genes that make up the MHC complex play important roles in immune responses. We can divide the MHC into the class I, II and III regions, each region containing groups of genes that code for proteins with distinct immune functions (Fig. 20).

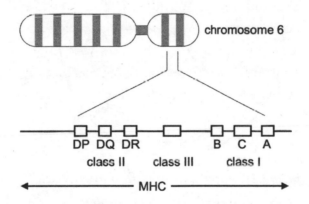

Figure 20. The major histocompatibility complex in man.

Class I MHC proteins are found on all nucleated cells – that is, virtually all cells. The function of MHC class I antigens can be illustrated by looking at what happens when a virus infects a human cell. The virus invades and multiplies inside the hapless cell, generating bits of viral protein, peptides that get associated with class I MHC proteins – coded by the MHC class I genes – within the cell. The MHC class I–viral peptide complex then makes its way to the cell surface, where it is displayed to a killer or cytotoxic T cell, which recognizes the peptide–MHC complex through its specific receptor (Fig. 21). Recognition leads to binding, and, once bound, the cytotoxic T cell destroys the infected cell. Foreign graft cells and some tumour antigens are also presented – by MHC class I antigens – to cytotoxic T cells, which set about engineering their destruction. It is no accident that class I MHC proteins are found on all nucleated cells: any cell that is foreign becomes infected or transformed into a malignant cell that should be capable of being destroyed by the immune system. No prisoners taken.

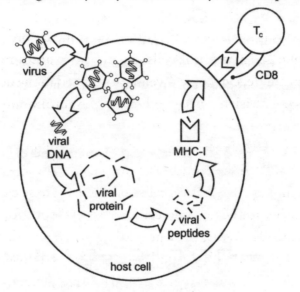

Figure 21. Antigen presentation to cytotoxic T cells by MHC class I protein.

An experiment that elucidated this neat killing function by T cells was carried out by Peter Doherty and Rolf Zinkernagel. We will come across this key experiment later in this chapter.

Let us first digress for a while and look at how a cytotoxic T cell kills a target cell. Two mechanisms are involved. One involves the release of proteins called 'perforins' and 'granzyme' from the T cell. The perforin molecules are inserted into the cell membrane of the target cell, making a plug with a hole through which granzymes can enter the target cells. Granzymes disrupt the cellular machinery, resulting in the death of the cell.

The second mechanism involves the T cell inducing the expression of death receptors called 'Fas' on infected cells. The T cell has a ligand for this receptor, a death activator designated 'Fas ligand' (FasL) on their surfaces, and as the T cell locks onto its target, FasL binds Fas on the surface of the target cell, leading to signals being generated inside the target cells. These signals cause the cell to commit suicide, a process called 'apoptosis'. Neat.

However, this is not a perfect system and some cytotoxic T cells can turn rogue and attack self-cells, giving rise to autoimmune disorders. For example, cytotoxic T cells are involved in the destruction of beta cells of the islets of Langerhans in the autoimmune disease type 1 diabetes.

In contrast, MHC class II proteins are only found on a specialized group of cells called 'antigen-presenting cells' (APCs) – examples include macrophages, dendritic cells and B cells. These capture and present antigens found outside cells – in the blood, for example – to T helper cells. But to present pathogen antigen, the antigen-presenting cell must engulf the pathogen and digest it inside the cell. Protein fragments from the pathogen, peptides, generated by this process can then bind with MHC class II proteins found in the cell interior. As before, the class II MHC–peptide complex gets transported to the cell surface, where the antigenic peptide is presented to a T helper cell, which, by secretion of cytokines, 'helps' B cells make antibodies, and turns resting cytotoxic cells into killers, thus unleashing the foot soldiers of the specific immune system against the pathogen (Fig. 22).[i]

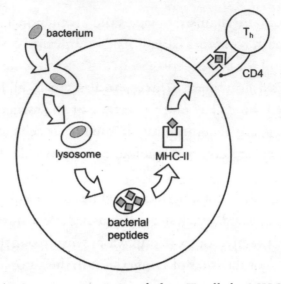

Figure 22. Antigen presentation to helper T cells by MHC class II protein.

Antigen presentation is the key function of MHC proteins. Antigens, once processed and presented to T cells by macrophages, dendritic cells and B cells, allow immune responses to be generated. Some of the early experiments that unravelled the mechanism involved in antigen presentation were carried out by the Cuban American immunologist Emil Unanue.

*

Unanue obtained his PhD from the University of Havana and migrated to the United States to pursue a career in immunological research. His major contributions to immunology began in the laboratory of Dr Brigitte Askonas at the National Institute for Medical Research at Mill Hill in London during the 1960s.

He made an interesting observation. When proteins were taken up and phagocytosed by macrophages, they became more immunogenic – that is, they led to a greater immune response.[9] This was an unexpected finding as it was generally believed that macrophages simply destroyed antigen, and that antigens needed to be intact molecules to be recognized by the immune system. What was going on? What were Metchnikoff's macrophages up to?

Unanue found that this enhanced immune response was directed towards antigen stuck to the cell membrane of the macrophage. Later experiments confirmed that the macrophage was presenting this surface antigen to T helper cells. He also showed that antigens were processed inside the phagocyte before being presented to T cells, and that these immunogenic fragments of antigen on the cell membrane of macrophages were peptides.[10]

In 1985, Unanue and his colleagues demonstrated, using radiolabelled peptides, the binding of internalized peptides to class II MHC molecules. This preceded the export of the specific peptide–MHC II complex to the cell surface for presentation to a T helper cell.[11]

The pathogens that trigger this helper T-cell response are typically extracellular, that is, those that live outside cells, like the pneumococcus that causes pneumonia. Therefore, MHC class II antigens are involved in the generation of immune responses to extracellular infection.

Baruj Benacerraf subsequently found that his immune-response genes were encoding MHC class II antigens found on macrophages and B cells and these antigens were capable of stimulating T cells *in vitro*. They were also found to be responsible for graft-versus-host reactions and graft rejection.[12]

Jean Dausset, George Snell and Baruj Benacerraf were jointly awarded the 1980 Nobel Prize in Physiology or Medicine. In his acceptance lecture, Dausset states:

Here again we find the *bipolar* division of the functions of the products of the HLA complex:

    1) Class I products appear to serve as targets when a cell is either infected by a virus or covered with a hapten.

2) Class II products appear to serve as a regulator between the various cell subgroups involved in the immune response.

In both cases, a phenomenon of restriction is most often observed, that is to say, an identity with class I or II products is apparently necessary between the cooperating cells.[13]

What is this phenomenon of 'restriction' that Dausset is referring to? The presentation of antigen by MHC class I and II antigens to T cells follows a rule called 'MHC restriction', a fundamental property of cellular immunity discovered in the early 1970s by Peter Doherty and Rolf Zinkernagel.

*

Doherty was an Australian veterinary surgeon with a PhD from Edinburgh University, Zinkernagel a Swiss immunologist who had earned his doctorate from the Australian National University. Together they set about investigating how cytotoxic T cells destroyed infected cells. In their now-famous experiment, Doherty and Zinkernagel used two strains of mice, strain X and strain Y, and studied how T cells isolated from these mice destroyed X and Y cells infected with lymphocytic choriomeningitis (LCM) virus. Here we go, another somewhat complex but essential experiment that must be described!

Culturing LCM-infected strain-X mouse cells with strain-X T cells resulted in the destruction of the infected cells by the T cells. However, when they added T cells from a strain-X mouse to LCM-virus-infected strain-Y mouse cells, the strain-X T cells failed to destroy the strain-Y cells. When the experiment was repeated using T cells from a strain-Y mouse, the strain-Y cells infected by the virus were destroyed (Fig. 23).[14]

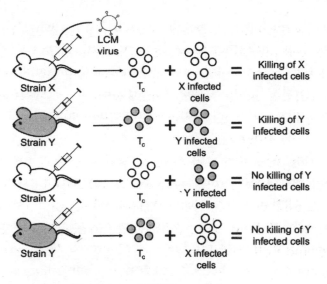

Figure 23. MHC restriction.

Doherty and Zinkernagel concluded that the cytotoxic T cells needed to recognize two antigens that were present on the infected cell: a viral antigen and an MHC antigen. Unless the T cell recognized both antigens, it could not attack and destroy the infected cell. Also, a T cell from an individual animal was *restricted* to interacting with a certain MHC, one which was uniquely found on cells belonging to the same individual. This property of T cells was called MHC restriction.

The cytotoxic T cell was subsequently found to bind to viral antigen lying in a specific cleft of the MHC molecule.[ii] In an interview, Peter Doherty explained the significance of their discoveries.

The fact that cytotoxic T lymphocytes – the killer T cells – cannot usually recognise foreign antigens unless these antigens are paired with MHC antigens is pretty fundamental...

It shows us that the immune system can recognise a third state – altered self – as well as self and non-self. When a virus has infected a cell and the cell is displaying viral antigens in addition to its own, it has become altered self. That's what's recognised and dealt with, rather than the viral antigens *per se*. The point is that the body treats altered self in much the same way as non-self. A virally modified cell is destroyed in the same way that a transplanted cell from another individual would be...

It gives us a biological role for the MHC system. People were wondering why the body should have a system for combating transplanted tissue when this state clearly never arises in nature. We suggested that the recognition of alloantigens – MHC antigens differing from your own – was there not to frustrate transplant surgeons but to help the body 'see' altered self... Altered self-recognition allows the body to conduct surveillance on its own cells... A cell's antigens can be changed not just in virus infections but in certain cancers, for instance.[15]

For their discoveries concerning 'the specificity of the cell-mediated immune defence' Rolf Zinkernagel and Peter Doherty were jointly awarded the Nobel Prize in Physiology or Medicine in 1996.

But these class I and II MHC antigens cause problems for transplant surgeons! The genes encoding them are highly variable, and therefore there are many different tissue types in a population. An individual is tolerant to its own HLAs, but if foreign HLAs are detected then the cells displaying them are attacked and destroyed by the immune system. Unfortunately, this means that transplants from donors that have not been matched for tissue type will be rejected.

Incidentally, there is another region, called class III, within the HLA. The class III region is not involved in defining tissue type and contains genes coding for several types of protein molecule involved in immune responses. These include complement proteins and several cytokines.

The search for a compatible transplant donor can be a long arduous process. The first crucial requirement is that the donor and recipient be blood-group compatible because ABO blood-group incompatibility leads to rapid rejection. Then we look at the HLA haplotype, the set of genes in the MHC complex between donor and recipient.[iii] A perfect match is not always possible or necessary. Siblings and parents of the recipient are tissue-typed first as there is a stronger likelihood of a match in one of these close relatives. Interestingly, class II antigens appear to be more important than class I antigens in eliciting rejection.

If there is no HLA-matched sibling or parent, then extended family members (aunts, uncles, cousins, and so on) can be HLA-typed. If all these come up as unsuitable, the search is extended to look for an unrelated donor. Such a search is performed through national and international transplant registries which have thousands or, in some cases, millions of potential HLA-typed donors.

*

Layla never contracted HIV. The resistance of some Pumwani women appears to be down to having certain types of HLA antigens on their infected cells.

These HLA proteins may be preferentially binding to parts of the viral proteins that are essential for the virus. These regions are 'conserved' and do not change with time. The infected cells can therefore be destroyed by cytotoxic T cells. In contrast, cytotoxic T cells in individuals with different HLA antigens that preferentially bind to HIV antigens that are more likely to mutate will struggle to recognize cells infected by the virus which has mutated. They will not be able to keep up with a changing virus. Studies have also shown that one such HLA class I antigen bound a conserved region of the p24 HIV capsid protein and was associated with increased survival in HIV-infected people: these were the 'elite suppressors'. These individuals can get infected but survive for decades without anti-HIV drugs. Their T-cell counts don't drop, and there is no evidence of viral replication. Conversely, having different types of

HLA class I proteins is associated with rapid disease progression to AIDS in some HIV-infected people.[16]

In the late 1960s and 1970s, the possession of certain variants of these gene sets, HLA haplotypes, were found to be associated with an increased susceptibility to certain autoimmune diseases. For example, the HLA B27 haplotype conferred increased susceptibility to ankylosing spondylitis, whereas HLA DR3 was linked to systemic lupus erythematosus and coeliac disease, and the possession of HLA DR4 made it more likely that individuals would suffer from the autoimmune disease type 1 diabetes.

i   Exogenous antigens (inhaled, ingested or injected) are taken up by antigen-presenting cells (APCs) by a process called endocytosis. The antigen is contained within an endosome, which fuses with lysosomes containing enzymes and various antimicrobial substances. Within the endosome, the antigen is degraded into fragments (such as short peptides). These antigenic peptides combine with MHC class II molecules and the MHC antigen complex is exported to the surface of the cell, where the antigenic peptide – nested within a groove of the MHC molecule – is recognized by a specific receptor complex on $CD4^+$ T cells.

CD4, a co-receptor of the TcR complex, binds to a different part of the MHC molecule. To become activated, a T cell needs two signals. The first is initiated by the binding of the peptide–MHC complex to the TcR–CD3 complex. This is followed by a second signal: binding of CD28 on the CD4$^+$ T cell to CD80 (B7-1) or CD86 (B7-2) on the APC. If there is no second signal, the T cell presumes that what is being presented is a self-antigen and does not respond. Once a naive T cell receives both signals it becomes activated. Once activated, the T helper cell proliferates by releasing and responding to the T-cell growth factor, interleukin 2 (IL-2). These activated T cells then become T helper 0 cells (Th0), which, depending on the cytokine environment present, further differentiate into the different T-cell subsets: Th1, Th2, Th3, Th17 or Tfh.

After many rounds of cell division, the Th cells differentiate into 'effector' and 'memory' Th cells. When antigen is re-encountered, the responding memory cells that generate the secondary response only require the first signal for activation.

The production of IL-2 by helper T cells is also necessary for the proliferation of activated CD8$^+$ T cells. Without help from CD4 cells, CD8$^+$ T cells do not proliferate.

ii Activation of cytotoxic T cells also requires two signals: TcR interaction with peptide-bound MHC class I molecule and a second signal provided by the CD28 molecule on the T cell binding with either CD80 or CD86 (also called B7-1 and B7-2) on the target cell.

iii Of the many HLA antigens, three, A, B and DR provoke the most vigorous anti-graft immune responses. As such, matching for these antigens is paramount for successful transplantation. As three alleles – coding for A, B and DR – are inherited from each parent, there are six alleles in total that need to be considered when HLA matching is carried out between donors and recipients. A perfect match would therefore involve both donor and recipient being identical for these six antigens (2 × A, B, DR): this is called a 6/6 match. However, matches are now being defined to a level of 10/10 on loci A, B, Cw, DR and DQ. Although clearly desirable, a perfect match is not always necessary for successful graft survival.

Several methods can be used to carry out HLA typing. In an older serological method, specific anti-HLA antibodies are used to detect the presence of these antigens on donor and recipient white blood cells. A more direct and sensitive method involving DNA sequencing enables identification of HLA genes coding for key transplantation antigens. Although serologic testing may be sufficient to identify a compatible sibling donor, more sensitive DNA sequencing methods are needed for the identification of compatible 'unrelated' donors. Molecular typing, being more sensitive and accurate than serology, can detect even small HLA differences between recipients and donors.

A crossmatch test is routinely performed prior to transplantation. In this test, a small amount of the recipient's serum is mixed with the potential donor's white cells. Here we are looking for the presence of recipient antibodies that might react with antigens on donor cells. If the recipient has antibody to donor HLA antigens, the donor cells will be lysed when complement is added. This is referred to as a 'positive crossmatch' and is a contraindication to transplantation because of the risk of rapid hyperacute rejection.

Chapter 28

THE CONDUCTOR OF THE ORCHESTRA

Missouri, 1968. On a warm spring day, a young African American boy walked into the Barnes Hospital in St Louis. Speaking with difficulty, he told the medical staff that he was aged 16 and his name was Robert Rayford. He appeared exhausted and, after staggering around unsteadily, collapsed onto a trolley. He was short of breath and was struggling to speak, but he managed to convey to the staff that he had been unwell for several months. History revealed little: the boy, from a poor black neighbourhood, had never been abroad or ventured far from his hometown. He had no significant medical history and had been well until his current illness. The examining physician noted that Rayford was thin, pale and covered in warts and sores. He had severe thrush inside his mouth and the lymph glands in his neck were swollen. The doctor also noted that his testicles and the area around the bottom of the spine were swollen. Rayford refused a rectal exam.

The doctors thought he was suffering from lymphoedema – swollen lymph channels – which was causing the swelling in his testicles and pelvis. They tried to drain the fluid but were unsuccessful. Meanwhile, through it all Rayford remained passive and seemed almost resigned to whatever fate had in store for him. 'He barely said boo,' one of his attending physicians, Dr Memory Elvin-Lewis, recalled later.[1]

Rayford was treated with antibiotics and initially appeared to respond, his condition stabilizing. But in March that year his symptoms returned, and he took a turn for the worse. His temperature spiked; he developed a cough and was diagnosed with pneumonia. His routine blood examination revealed a drastically reduced white-cell count. As we know, white blood cells make up our defence force, and a reduction in their numbers makes a person immunodeficient: unable to defend themselves against pathogens and certain cancers.

On 15 May 1969, some 15 months after presenting himself at the hospital, Robert Rayford succumbed to his mysterious illness and died. The official cause of death was bronchopneumonia, but the constellation of symptoms indicated a hitherto unknown disease.

Dr William Drake carried out the autopsy. He noted a small purple lesion on Rayford's left thigh, subsequently finding similar lesions in the boy's rectum and anus and, upon internal examination, in several organs. The lesions were later confirmed to be Kaposi's sarcoma, a rare type of skin cancer that was usually only found in elderly Mediterranean men, Ashkenazi Jews or individuals who had a severe underlying immune deficiency. Finding it in a 16-year-old boy was highly unusual. He also had severe proctitis: inflammation of the anal canal. Dr Drake took some blood and tissue samples and sent them away to be stored frozen in a facility at the University of Arizona. That concluded the case of Robert Rayford, and, like the countless poor souls who shuffled their way through the hospital, recovering or dying of whatever condition that had brought them in, he was forgotten.

Just over a decade later, between October 1980 and May 1981, five young homosexual men aged between 29 and 36 were admitted to three different hospitals in Los Angeles, California. The men did not know one another and had never met, but they all had a remarkably similar set of symptoms. They were suffering from severe oral thrush, candida infection, were short of breath and coughing, and had a fever. Laboratory investigations revealed a decreased white-blood-cell count and evidence of being infected with cytomegalovirus, a rare infection that was only seen in cases where the immune system was greatly compromised. The men also had another unusual problem, a chest infection caused by *Pneumocystis jirovecii*, called *Pneumocystis carinii* at the time, an opportunistic parasite that was hardly ever found to cause disease in healthy individuals but, once again, was often seen in the immunodeficient.

When white blood cells from three of the five patients were placed in a dish and mixed with an antigen, the cells showed a decreased ability to divide.

The other two patients were not given this test, but by 1984 it had been confirmed that the strange illness that these five young men had contracted was AIDS – 'acquired immune deficiency syndrome' – caused by a virus later named the 'human immunodeficiency virus', or HIV.

The reports of these early AIDS cases got one of the doctors who had treated Rayford 15 years earlier thinking. Dr Marlys Witte of the University of Arizona, who had been present at the Barnes Hospital back in 1969 and had been part of the team of doctors who had tried to save the boy, noticed similarities between Rayford's case and those of the five young men from Los Angeles. She contacted her former colleague Dr Elvin-Lewis, and in 1987 they decided to thaw out Rayford's blood and tissue samples and test them for the presence of HIV. Rayford's blood sample showed the presence of antibodies against all nine HIV proteins, revealed in a test called a 'western blot'. Later, another viral protein, the 'p24', was detected in the samples. There was no doubt that Robert Rayford had HIV and probably died because of AIDS.

But how had a teenager from the black ghetto in Missouri contracted the disease? There was no record of Rayford ever having received a blood transfusion. He had denied being homosexual, even though the presence of severe proctitis suggested otherwise. So where had he picked up the infection? He had never travelled abroad or been to any of the cities, such as Los Angeles, where the disease was now spreading with speed through the gay population. It was a mystery – one that, to date, has not been solved. Robert Rayford was, however, the first person in the United States who had been shown to have been exposed to HIV, and someone who, in all likelihood, had died from AIDS way back in 1969.

To date, the AIDS epidemic has caused an estimated 35 million deaths worldwide and in 2015 nearly 37 million people were living with the disease.[2] AIDS develops when the human immunodeficiency virus infects one of the key cells of the immune system, the T helper cell – referred to as the conductor of the immunological orchestra – resulting in the disabling of the immune system, which leaves the victim susceptible to a range of infections and cancers. This highlights the importance of the immune system and the devastating consequences of its loss to human health and survival.

*

The humoralists and the cellularists battled for five decades, Ehrlich and his tribe squaring up against Metchnikoff and his followers. The pendulum swung backwards and forwards, each camp insisting on the primacy of their component: the glorious antibody or the hungry white blood cell, fighting the ubiquitous pathogen.

Now we know that they are both important and interdependent. They are part of a complex orchestra that works in concert to battle the enemy.

But who is the conductor of this immunological orchestra? Who sets the tempo? Who controls all the different elements: antibodies, lymphocytes, cytokines, macrophages? Who coordinates the ensuing responses that will neutralize pathogens? And what happens if this conductor is targeted by a pathogen and disabled? What would happen to the orchestra, to the immune system?

We now know that the conductor is the T helper cell. It sits at the centre of the complex immune network, helping B cells produce antibody, activating cytotoxic T cells and macrophages – it gives the orders (Fig. 24).

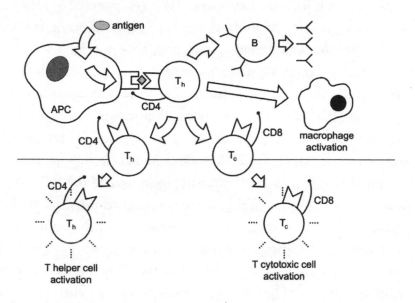

Figure 24. The conductor of the immunological orchestra.

After millions of years of evolution, inevitably perhaps, nature produced the ultimate pathogen. From the darkness of a distant forest a virus emerged. A parasite of a wild monkey, it jumped species and found a new host, one that enabled it to spread with a terrifying alacrity. What was most frightening about this virus was its target: the T helper cell, the conductor of the immunological orchestra. Like a sniper sitting atop a high building, it picked off these crucial cells one by one.

Robert Rayford, the boy from Missouri who died in 1969, was attacked – we don't quite know how or when – by the virus, which targeted his T helper cells. The disease that ensued was called AIDS.

The human immunodeficiency virus paralysed the victim's immune system. Once that was made ineffective, it left the victim open to the pathogens and cancers that were ordinarily controlled. The spread of AIDS highlighted the importance of the T helper cell. It also showed us what happens to a person when the T helper cell population is slowly but inexorably decimated.

Let's see what happens to someone who gets infected by HIV. First, the virus enters its victim. In the case of HIV infection, transmission occurs through sexual intercourse, by a contaminated blood transfusion or vertically from mother to baby. Once in the bloodstream, HIV, through a specific surface protein receptor that binds a molecule (CD4) on the T helper cell, attaches itself to a T cell. This specific viral protein is the key that fits the CD4 receptor (the lock) on the T helper cell. It then sinks into the cell. Sometimes the virus makes copies of itself, then explodes out of the cell to infect others; sometimes the viral genome is surreptitiously inserted into the host genome, waiting, like a sleeper agent, for the right moment to emerge and multiply, wreaking destruction in its wake. Either way, immediately following infection the T helper cell numbers begin to plummet. The immune system puts up a fight; cytotoxic T cells attempt to hold the virus in check and for a while a stalemate ensues – the T helper cell numbers remain low but stable. But it is only a matter of time, and if untreated, the virus starts to get the upper hand.

As the T helper cell numbers decline, the victim is left open to opportunistic infections and certain cancers, again highlighting the importance of the T helper cell in affording protection against these diseases. Healthy individuals have a T helper ($CD4^+$) cell count between 500 and 1,500 cells per cubic millimetre of blood. As long as your $CD4^+$ cell number is not less than 500 you are protected, but once it dips below that figure then the importance of a functioning immune system, and more particularly the importance of the $CD4^+$ cell, becomes apparent.

When the $CD4^+$ cell count falls to between 500 and 200 cells per cubic millimetre, patients begin to experience infections, but these are generally minor, such as thrush, shingles and sinusitis. Oral thrush, or candidiasis, manifests as white patches on the tongue, gums or the lining of the mouth, and some patients may report difficulty in swallowing. If the candida infection is in the vagina, a thick, creamy discharge is often seen, along with itching and redness.

More alarmingly, lesions of a cancer – Kaposi's sarcoma – may appear as the count drops below 500. These purplish marks are caused by the human herpesvirus 8, and, before the introduction of anti-HIV medication, around 20 per cent of AIDS patients presented with them.[3] The lesions can be anywhere: on the skin or mucous membranes, or in the internal organs. If you recall, these lesions were found during Robert Rayford's autopsy.

The key number of T helper cells that you must have to keep serious diseases at bay is around 200. Once the $CD4^+$ count drops any lower, an entire range of dangerous opportunistic infections and cancers make their grim appearance. Thus 200 $CD4^+$ cells per cubic millimetre of blood is considered a significant cut-off point: any further decrease significantly impacts on the ability of the immune system to carry out its defence functions.

Infections and cancers manifesting at this stage are called 'AIDS-defining conditions', and the patient now has AIDS. *Pneumocystis jirovecii* makes its appearance once the count drops below 200. This yeast-like fungus, which causes shortness of breath, dry cough and fever, is a common cause of death in HIV-infected patients. Another fungal infection that is seen in these circumstances is 'histoplasmosis'. This is an infection that spreads, and can present with fever, fatigue, chest pain, cough and headache, the exact symptoms depending on the location of the pathogen.

The importance of having an intact immune system is highlighted by a viral infection that causes disease once the $CD4^+$ count falls below 200. Usually kept in check by the immune system, the John Cunningham (JC) virus causes a devastating neurological disease called 'progressive multifocal leukoencephalopathy' (PML) in immunocompromised patients. Patients suffer from seizures, confusion, dementia, speech difficulties and many neurological signs. There is no definitive treatment for this condition and the mortality rate can approach 50 per cent.[4]

An infection that makes its appearance when the CD4 count is in the 200-to-100 range is cryptosporidiosis. This is a parasitic gastrointestinal infection in which patients suffer from chronic diarrhoea, vomiting and crampy abdominal pain. The infection is generally transmitted through contact with contaminated water, and the parasite has been isolated from swimming pools, lakes and – rarely – public water supplies. The infection can also be acquired by eating uncooked shellfish such as oysters. Person-to-person transmission through handling faeces from infected individuals can also occur.

A fungal infection, cryptococcosis, is also seen when the $CD4^+$ count is in this range. This infection, typically acquired by inhalation, is caused by the yeast *Cryptococcus neoformans*. It can then spread to the brain, causing meningitis, with patients presenting with symptoms such as headache, neck stiffness, fever, personality changes and loss of memory.

An asymptomatic viral infection that most individuals would have unknowingly encountered by age 40 causes disease when the $CD4^+$ count drops below 100. Cytomegalovirus (CMV) infection, transmitted through saliva, blood or semen, can cause inflammation of the back of the eye – the retina – in HIV patients. It can also cause gastrointestinal-tract infection, causing abdominal pain, diarrhoea and painful swallowing.

Once the $CD4^+$ count drops below 100, the HIV patient can also develop non-Hodgkin lymphoma, a cancer of immune cells. This usually presents as a painless swelling of a lymph node in the neck, armpit or groin. Symptoms, which depend on the location of the tumour, include bone pain, weight loss, night sweats and abdominal pain. More opportunistic infections also appear, often with an alarming rapidity. Patients can be infected by a parasite called

Toxoplasma gondii. Usually carried by animals such as pigs, cats and birds, toxoplasma can be found in contaminated pork and soil, as well as in cat litter. It can cause inflammation of the brain, causing headache, confusion, seizures and fever. Neurological signs such as weakness may also be observed.

AIDS patients encounter *Mycobacterium avium*-complex infections once the $CD4^+$ count drops below 50 cells per cubic millimetre of blood. These bacteria, common inhabitants of soil and water, can cause lung or gut infections, which in some instances can spread throughout the body, causing death. Patients present with fever, night sweats, fatigue, diarrhoea and abdominal pain.

Primary brain lymphoma also appears when the $CD4^+$ cell count drops below 50. These cancers cause headaches and focal neurological signs such as weakness or sensory loss. Patients can also suffer from seizures, confusion and changes in personality.

Women who are infected with HIV are more likely to suffer from invasive cervical cancer caused by the human papilloma virus (HPV). Symptoms include post-coital bleeding, passing blood in the urine, post-menopausal bleeding, low back pain, weight loss and, following the spread of the disease, bone pain. A recent study showed that, compared to uninfected women, HIV-infected women with $CD4^+$ T-cell counts of above 350, 200 to 349 and below 200 cells per cubic millimetre had a 2.3, 3.0 and 7.7 times increase in the incidence of invasive cervical cancer, respectively.[5]

All these grim findings point to one conclusion. An intact immune system is essential for survival. Without enough T helper cells, antibody and cellular immune responses are compromised, and the individual is open to the relentless assault of pathogens and cancer.

Chapter 29

THE BEAUTIFUL CELL

The 50-year-old dispute between Paul Ehrlich and Ilya Metchnikoff was finally laid to rest by the discovery of a new cell, one that was shown to act as a bridge between cellular and humoral immune responses. Universally recognized as a key initiator of immune responses, it was shown to be intimately involved in both innate and specific immunity. This cell was discovered by the Canadian immunologist Ralph Marvin Steinman.

Steinman, of Ashkenazi Jewish descent, was born in 1943 in Montreal. He received a Bachelor of Science from McGill University followed by an MD, *magna cum laude*, in 1968, from Harvard Medical School. In 1970, Steinman began his research career at the Rockefeller University in the laboratory of the renowned microbiologist René Dubos.

By the late 1960s, it had become possible to study cell-to-cell interactions *in vitro*. Antigens and immune cells could be added to a dish and the proliferation of immune cells, indicating a response to the antigen, could be measured. It was also possible to quantify antibody-producing cells by using the plaque-forming cell assay that we came across in Chapter 22.

Simply mixing T and B cells together with antigen did not necessarily lead to antibody production by the B cell. The *in vitro* assays indicated that another cell, an accessory cell, was also required. At the time, this accessory cell was believed to be a macrophage, which, by taking up and presenting antigen to a T cell, allowed the latter to help the B cell make antibody. In fact, as we saw in Chapter 27,

Unanue and others had shown in the 1970s that macrophages, when loaded with antigen, could present antigenic peptides to T cells, leading to the initiation of immune responses.[1]

Steinman's early experiments showed that macrophages were not the only cells involved in antigen presentation. There was another cell that was more efficient at presenting antigen to T cells. Peering at a preparation of mouse spleen cells through a phase-contrast microscope, Steinman found the cell that he was looking for. The cell had long outgrowths – known as 'processes' – like the thin branches of a tree. Electron microscopy showed the cell in more detail. It was constantly extending and retracting its long processes (Fig. 25).

Steinman first announced the discovery of his cell at a meeting in Leiden in the Netherlands in 1973. He told the audience that the long appendages of the cells reminded him of his tall and graceful wife, Claudia. 'I thought about calling them claudiacytes,' he said.[2] In the end Steinman called his discovery 'dendritic cells', from the Greek *déndron*, meaning 'tree'. He had discovered a new cell, one intimately involved in immune responses.

Figure 25. Dendritic cell.

After managing to isolate and purify dendritic cells, Steinman was ready to investigate their functions. Even though these cells constituted only 1 to 2 per cent of all cells found in the spleen, they were over a hundred times more efficient at causing T-cell proliferation than any of the other cells in the spleen-cell mixture.[3] Proliferation was quantified by measuring the amount of radiolabelled thymidine (the nucleoside T, which pairs with A in DNA) taken up by the dividing cells, the thymidine being incorporated into the new DNA synthesized in the progeny cells. So a higher incorporation of thymidine indicated more synthesis of new DNA molecules, which in turn indicated more cell multiplication.

These initial findings did not receive universal acceptance, and the response to Steinman's findings was somewhat muted: the macrophage was still considered to be the key cell that presents antigen to T cells. Metchnikoff's shadow was long and enduring.

Then, in 1980, using haptens as antigen, Steinman showed that dendritic cells were able to present antigen to cytotoxic T cells.[4] So the strange cell with the long processes was capable of presenting antigen to both helper and cytotoxic T cells (Fig. 26). And by presenting antigen to the helper T cell, the dendritic cell played a role, albeit indirectly, in antibody production.

This feature of the dendritic cell – the ability to present antigen to both T helper cells and T cytotoxic cells – is called 'cross-presentation'. How was the dendritic cell able to carry this out? Later experiments showed that the dendritic cell was able to take up antigens, process them, and present them through both MHC class I and class II molecules to cytotoxic and helper T cells, respectively.[5] In this respect, Steinman's cell, by having both class I and class II molecules on its surface, was acting as both a traditional antigen-presenting cell and any other infected or altered cell. Therefore, the dendritic cell is central to the clonal selection theory as it is the main cell that takes up antigen, and by selecting from the clones of T cells presents the antigen to a unique T cell, which will then proliferate.

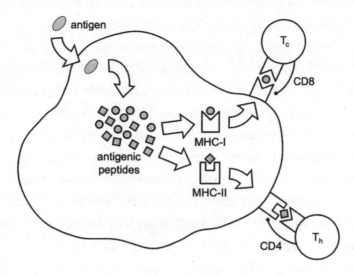

Figure 26. Cross-presentation of antigen by a dendritic cell.

But where were these dendritic cells found? Steinman discovered them residing at mucosal surfaces, the inner linings of the respiratory, gastrointestinal and urinogenital tracts. Here they were ideally placed to act as sentinels, able to capture antigens from the environment and present them to T cells, generating specific immune responses: the dendritic cells were our guardian cells, part of the border patrol capturing illegal aliens and then presenting them for processing and deportation.

Additionally, Steinman found dendritic cells in lymphoid organs including the lymph nodes, the spleen and even the thymus.[6] These sentinel dendritic cells were thus able to present antigens to the lymphocytes when they, as James Gowans had discovered, weaved their way in and out of lymphoid tissues.

Steinman continued studying the dendritic cell, finding in 1985 that the cell required activation before it was able to present antigen.[7] This activation was achieved by signals the dendritic cells received from various receptors normally found on pathogens. When activated, the dendritic cells increased their expression of

MHC molecules, leading to enhanced antigen presentation with consequent initiation of immune responses.

Around 2000 and 2001, Steinman discovered an unexpected role played by the dendritic cell. They were also involved in tolerance, and, as before, it all seemed to depend on whether the dendritic cell got the right signals from a pathogen.[8] If it did, it was activated and presented pathogen antigens to T cells, and an immune response ensued. If, however, the dendritic cells were 'resting' and not receiving pathogen signals, immune responses were not elicited. That seemed reasonable, but it turned out that the dendritic cells were not really 'resting': they were busy doing something else, something quite unexpected.

These 'resting' dendritic cells, Steinman and his colleagues found, were continually capturing self-antigens from cells that were dying. These antigens were then presented to T cells, but because the pathogen-related signals were absent, the T cells, potentially capable of reacting to self-antigen, became 'anergized': that is, they became unresponsive, or were deleted or transformed into regulatory T cells. As such T cells were unable to generate immune responses, the dendritic cell was preventing harmful immune responses being potentially mounted against self-antigens. So the dendritic cell played a significant role in tolerance, a fundamental feature of the immune system, elegantly articulated by Peter Medawar back in the 1940s.

A wealth of data began to accumulate: there was no doubt that the dendritic cell played a crucial role in immunity. But, Steinman mused, was there a practical application here? Could the dendritic cell be used therapeutically? The creation of a novel vaccine, perhaps? But as he planned more experiments, ideas racing through his mind, Ralph Steinman had no idea of the unexpected twist that was to unfold in this remarkable scientific story.

In one experiment, Steinman added tumour antigens into a suspension of purified mouse dendritic cells placed in a dish. The idea was to 'load' the dendritic cells with tumour antigen, and, once this was achieved, to reinfuse the cells back into a mouse. When challenged with tumour cells, these mice, to Steinman's delight, mounted vigorous immune responses against the tumour. The antigen-loaded dendritic cells were activating large numbers of tumour-antigen-specific T cells. Repeating the experiment with viral antigen-loaded dendritic cells gave the same result. So this method of immunizing animals with antigen-loaded dendritic cells appeared to afford protection against viral infections and tumours.[9] Could this be a completely new vaccine strategy, Steinman wondered, a type of immunotherapy that could ultimately be used in patients?

In a cancer patient, this would first involve extraction of tumour antigen from a surgically excised tumour. These antigens would then be mixed outside the body or *ex vivo* with dendritic cells isolated and purified from the same patient. The loaded dendritic cells would then be reintroduced back into the patient, and, if all went well, would present tumour peptides to helper and cytotoxic T cells, which would seek and destroy any tumour cells that were still lurking inside the patient.

Steinman attempted to load dendritic cells with a variety of antigens prepared from several different cancers and pathogens including *Mycobacterium tuberculosis*. He tried using specific antibodies raised against tumour or pathogen antigens to get these antigens into the dendritic cell. These antibodies, carrying their specific antigens, bound receptors on dendritic cells – ones that bound the Fc end of the antibody molecule – and were taken up by dendritic cells.[10] So the specific antibodies were delivery systems that could transport antigens into dendritic cells.

Then events took an unexpected turn. In the spring of 2007, Ralph Steinman, then aged 64, returned home from a skiing holiday suffering from a severe bout of diarrhoea. A few days later he developed jaundice and shortly afterwards received the worst possible news. He had pancreatic cancer, and it had spread to his lymph nodes: it was advanced, stage four. The one-year survival rate for stage-four pancreatic cancer was grim. In fact, 80 per cent of patients were dead within the first year, and 90 per cent the year after. He told his family about the diagnosis but advised them not to Google it.[11]

The tumour was still amenable to surgical resection, and, on 3 April 2007, two weeks after his diagnosis, Steinman went in for surgery. He had already made plans for his tumour. Very soon after his diagnosis, Steinman had called a meeting with two former colleagues of his laboratory: Leon Rosenberg from the Rockefeller and Ira Mellman from the gene-technology company Genentech. 'It was the weirdest experience,' Mellman says. 'It was like we were having a lab meeting from the old days: talking about what experiments to do, what needed to be found out, how interesting it was what you can and can't do. It was totally natural scientific discussion except we were talking about his tumour.'[12]

The four-hour procedure to remove Steinman's tumour took place at the Memorial Sloan Kettering Cancer Center, a few blocks from his laboratory. The surgeons removed the 2½-inch-long growth and gave it to Dr Sarah Schlesinger, a friend and colleague of Steinman, who also happened to be a pathologist at the Sloan Kettering at the time.

Teams of researchers set to work on the tumour. A cell line was produced from bits of tumour grown in mice. The tumour-cell cultures were then subjected to all the available chemotherapeutic drugs and the responses were assessed. DNA from the tumour was sequenced to look for mutations that might be targeted by specific drugs. Peptides were extracted from the surface of tumour cells to prepare vaccines.

After surgery, Steinman continued with the standard chemotherapeutic treatment that was offered, but his thoughts, as always, were on his beautiful cell. Would this cell, the one he had discovered and studied for decades, be his saviour? Would the therapeutic vaccines he could make from them save him from what seemed to be his inevitable fate?

Steinman tried eight types of treatment. According to Jedd Wolchok, a medical oncologist at the Sloan Kettering, 'It was the ultimate experience in personalised medicine.'[13] But there were no shortcuts, no secret injections at the back of the lab. Steinman and his colleagues submitted all these compassionate-use protocols for approval by the US Food and Drug Administration (FDA) before proceeding.

One vaccine he tried, called 'GVAX', used irradiated cells from his tumour. A gene was inserted into the growth, one that produced a cytokine that, when secreted by the engineered tumour cell, caused dendritic cells to move towards the tumour.

The second approach involved loading dendritic cells isolated from Steinman with RNA extracted from his tumour. The tumour RNA, once inside the dendritic cells, would produce tumour proteins, antigens that would end up being displayed on the surface of the dendritic cells. And when these dendritic cells were injected back into Steinman they would – hopefully – present these tumour antigens to T cells, leading to an effective anti-tumour response: one that would seek and destroy any tumour cells bearing these antigens on their surfaces.

The third vaccine, using a similar approach, involved adding peptide antigens extracted from Steinman's tumour cells to dendritic cells, again taken from Steinman. As before, these loaded dendritic cells were then injected back into Steinman in the hope that they would activate his T cells.[i]

In addition to these experimental vaccines, Steinman also tried several conventional, and even some novel, anti-tumour drugs that were around at the time. He monitored the progress of his tumour by measuring the levels of a tumour marker, CA 19-9, in his blood. The levels of this tumour marker increased when a tumour was growing and spreading and decreased when the tumour was responding to treatment.

Steinman's CA 19-9 levels were found to decrease after certain types of treatment, but because he was taking both treatments concurrently, it wasn't possible to say whether the response was due to the chemotherapy or to his vaccines. But, according to science writer Lauren Gravitz, who met Steinman and reported on his ordeal for *Nature*, he remained optimistic throughout these gruelling treatments.[14]

While all this was going on, Steinman's routine continued as before. He put in a full day in his lab, often working late into the evening. Then, back home, he would continue to work at his computer. He attended conferences all over the world and appeared to be doing well.

Four years went by, way past the survival time of most patients with stage-four pancreatic cancer. His initial prognosis had been for less than a year. It was impossible to know if it was his dendritic-cell vaccines that were prolonging his life. 'We knew at the onset that we wouldn't be able to tell which therapy made the difference,' Schlesinger remarked. 'We only had one patient so there is no statistical significance.'[15]

In September 2011, Alexis, Steinman's daughter, was asked by her mother to visit him. He was coughing violently and didn't look well. Steinman told her that the cancer had reached his bones. The CA 19-9 levels were also up.

Steinman spoke to Schlesinger just the day before he entered hospital for the final time. He discussed with her some data from a new HIV vaccine they were testing. They talked for several hours. 'I could see him getting sicker, but his spirit was so indomitable, and he was so optimistic,' she said.[16] But after that conversation Steinman's condition deteriorated rapidly. On Sunday, 25 September, Steinman became short of breath and was taken to hospital. He developed pneumonia followed by a pulmonary embolus: a blood clot in his lung. Ralph Steinman, the man who discovered the beautiful cell, died with his family around him on Friday, 30 September 2011. The family decided to inform the Rockefeller about his death on the Monday morning, which was 3 October.

On 3 October 2011, at 11.30 a.m. Central European Time, the Nobel Assembly in Stockholm, Sweden, announced the names of the 2011 laureates in Physiology or Medicine. The list of recipients included Ralph Steinman. They sent an email to Steinman informing him that he was to receive the Nobel Prize. At 5.30 that morning, Claudia Steinman heard Ralph's mobile phone buzzing, and found the email. Later that morning, the Rockefeller Institute, having now received the news of Steinman's death, formally notified the Nobel Assembly at the Karolinska Institute. It was 2.30 p.m. Central European Time: Steinman had passed away just three days before.

The press release by the organizers of the Nobel Prize read as follows:

The events that have occurred are unique and to the best of our knowledge are unprecedented in the history of the Nobel Prize. Considering this, the Board of the Nobel Foundation has held a meeting this afternoon.

According to the statutes of the Nobel Foundation, work produced by a person since deceased shall not be given an award. However, the statutes specify that if a person has been awarded a prize and has died before receiving it the prize may be presented.

An interpretation of the purpose of this rule leads to the conclusion that Ralph Steinman shall be awarded the 2011 Nobel Prize in Physiology or Medicine. The purpose of the above-mentioned rule is to make it clear that the Nobel Prize shall not deliberately be awarded posthumously. However, the decision to award the Nobel Prize to Ralph Steinman was made in good faith based on the assumption that the Nobel Laureate was alive. This was true – though not at the time of the decision – only a day or so previously. The Nobel Foundation does believe that what has occurred is more reminiscent of the example in the statues concerning a person who has been named as a Nobel Laureate and has died before the actual Nobel Prize award ceremony.

The decision made by the Nobel Assembly [i.e. to award Ralph Steinman the Nobel Prize] at Karolinska Institute thus remain unchanged.[17]

Speaking at the Nobel ceremony on behalf of his friend and colleague, Michel Nussenzweig said, 'Steinman's dendritic cell was the missing link that finally showed the interrelationship between the cellular and humoral immune responses. It finally solved the puzzle and bought the opposing theories of Metchnikoff and Ehrlich together.'[18]

---

i   Several vaccine experiments looked at tumour-specific antigen presentation by dendritic cells. In one approach, dendritic cells isolated from mice were cultured with tumour fragments and the cytokine GM-CSF. The cells were then reinfused into the mice. Strong anti-tumour immunity was observed when the mice were subsequently challenged with live tumour cells. Analysis revealed that that both helper and cytotoxic cells specific for the tumour antigens had been activated.

Steinman's work on dendritic-cell vaccines led to the approval, by the FDA, of the first dendritic-cell-based vaccine. On 29 April 2010, sipuleucel-T (sold under the brand name Provenge) was approved for the treatment of asymptomatic or minimally symptomatic metastatic hormone-refractory prostate cancer. This vaccine was designed to induce an immune response to a common prostate tumour antigen, prostatic acid phosphatase (PAP).

In this individual-based therapy, dendritic cells are first isolated from a prostate-cancer patient and then stimulated *ex vivo* by a fusion protein consisting of PAP and GM-CSF. These autologous dendritic cells, which are expanded as a result, are then reinfused into the patient. In trials, men treated with this vaccine showed survival increases of just over four months.

Another approach involves the transfection of tumour cells with the gene for GM-CSF. These engineered tumour cells, when reinfused into the patient, will, by the secretion of GM-CSF, enhance development and function of local dendritic cells drawn to the tumour. Such dendritic cells will then present tumour antigens to helper and cytotoxic T cells.

Summary

METCHNIKOFF'S FINGER

We can summarize what was learned in the golden age of immunology by using an example which involves the hero of cellular immunology, Ilya Metchnikoff. If you recall, the fiery Russian discovered phagocytosis by inserting a rose thorn into a starfish larva. Let us imagine that in his excitement Ilya pricked himself with the thorn. Cursing loudly – in Russian – our hero retreated. The thorn had had the effrontery to fight back!

Let's look at the events that unfolded. Microbes coating the thorn have now entered Metchnikoff's body, reaching deeper tissue by penetrating one of his most enduring innate defence barriers: his skin. There they will come face-to-face with the sentinel: dendritic cells. Certain receptors (pathogen-associated molecular patterns) on the microbes will bind receptors (toll-like receptors) on dendritic cells and they will be taken up by the dendritic cell. Inside the dendritic cell the pathogen is digested, and its peptides are processed and presented through class II MHC molecules to a specific T helper cell which has a receptor that binds the specific pathogen peptide – MHC complex.

Now, this T cell was born in the bone marrow and went to the thymus to complete its development. In the thymus the T cell underwent a rigorous selection process, in which any self-reactive T cells were eliminated. This is called 'central tolerance'. But this T cell passed the selection and was allowed to leave the thymus and enter secondary lymphoid organs to carry out effector functions against pathogens. Once antigen has been presented, the

T cell will proliferate and, by secreting cytokines, help B cells to make antibody.

Like the T cell, the B cell has, sitting on its surface, specific antibody receptors identical to the antibody the B cell would secrete. And again, like the T-cell repertoire, there are millions of B cells, each with a unique antibody receptor capable of secreting an antibody molecule with a unique specificity. How did these B cells achieve such diversity? It was down to a process of gene reshuffling, which rearranged and recombined different gene segments to create a unique antibody molecule. This mechanism was responsible for specificity and the generation of diversity, producing the wide range of antibody molecules and T-cell receptors and giving rise to a vast, limitless repertoire of immune cells and antibodies capable of reacting to anything that entered along with the rose thorn: anything nature could throw at us.

When activated, the T- or B-cell specific for antigen starts to divide, producing an identical clone of cells. This is clonal selection: only the clone with an antigen receptor specific for and binding to the antigen will proliferate. Some of these cells become effector cells of the primary response, while others – memory cells – lie in wait for a return when they unleash the heightened secondary response. This, if you recall, is the basis of vaccination. We have thus come across the hallmarks of specific immunity: specificity, diversity, self–non-self-discrimination and memory.

Antibodies produced by plasma cells would bind pathogens through the variable regions of heavy and light chains, and by activating complement make them more readily phagocytosable: opsonins. Some antibodies will neutralize toxins produced by the pathogens while others may physically block pathogens from attaching to epithelial cells.

T helper cells also activate cytotoxic T cells: these are the killers of the immune system. Any pathogens that have managed to get inside host cells would be broken down and their antigens presented to cytotoxic T cells through MHC class I antigen. The cytotoxic T cells then deliver the kiss of death, destroying the infected cell only if there is MHC identity between target and cytotoxic cell: MHC restriction. Cytokines secreted by helper and T cytotoxic cells also cause the destruction of the pathogen.

This is what was discovered during the golden period of immunology. There is of course a lot more that has been discovered since these early seminal discoveries, but there is also a lot more that remains unknown about the mysterious immune system.

PART THREE

PANDEMIC 2020:
THE LESSONS FROM THE PAST

Chapter 30

BACK TO 1890

Dongguan, China, 30 January 2020. A 69-year-old woman presented to her local hospital complaining of having a cough and feeling feverish for the past week. After testing positive for Sars-CoV-2, she was admitted to an isolation ward. She had a history of hypertension but was otherwise well. A CT scan of her chest revealed ground-glass shadows along her lung bases: she had pneumonia. She was started on antivirals and interferon alpha by inhalation.

At 4 p.m. on 4 February, her oxygen saturation dropped to 56 per cent. A chest X-ray showed that the inflammation in her right lung was spreading. She had developed ARDS and was taken to the intensive care unit, where she was placed on a ventilator. In addition to the antivirals, she was put on antibiotics and antifungal drugs as she was found to be co-infected with a fungus and bacteria. At 6.30 p.m. on 11 February, she went into septic shock, her blood pressure plummeting to 89/44. Bloody mucus was aspirated from her lungs; the haemoglobin count started to drop.

She was given 900 millilitres of matched plasma in three separate doses over the next two weeks. Her viral load started to decrease, and after spending nearly two months on a ventilator she started recovering and was extubated on 3 March. She tested negative for Sars-CoV-2 on 9 March. Chest X-rays showed improvement and she was discharged four days later.[1]

As this case was not part of a randomized controlled study one cannot conclude that the patient's recovery was down to receiving the plasma. But it was a possibility, especially as she had recovered from septic shock.

Another study documented the treatment of ten patients with convalescent plasma. These patients, who had severe Covid-19 disease, were given a single 200-millilitre dose of serum collected from recovered donors, most patients being treated around two weeks after infection. Following the infusion, their oxygen saturations, lymphocyte counts and X-ray findings improved, while levels of the inflammatory marker C-reactive protein (CRP) decreased. Virus was undetectable seven days after transfusion and, importantly, no adverse effects were noted. In this study, the ten patients who received serum therapy were compared historically to a control group consisting of ten patients matched for age, gender and disease severity. While baseline parameters of these two groups were similar, outcomes for the group treated with convalescent plasma were significantly different, with most patients well enough to be discharged. In contrast, only one person improved in the control group; three died and six stabilized. Even though the results were optimistic, large-scale studies showed no overall benefit of convalescent plasma therapy in Covid-19.[2]

*

These are examples of passive immunity. In 1892, Paul Ehrlich showed that protective antibodies can be transferred from mother to foetus through the placenta and, following birth, to the infant by suckling. This is an example of natural passive immunity: the recipient's immune system is not engaged. Behring's horse-antitoxin therapy, which prevented the deaths of thousands from diphtheria and tetanus, was an example of artificial passive immunization. This afforded protection but, again, the recipients relied on antibodies supplied by the animal to neutralize the toxins. But adverse reactions were also observed. As we saw in Chapter 17, Pirquet had observed 'serum sickness' in some of his patients.

These were due to a type III hypersensitivity reaction to the foreign antibody.

In addition to using animal serum, Behring and others used convalescent human whole blood or sera to treat diphtheria and tetanus.[3] Convalescent plasma was also used during the Spanish-flu pandemic in 1918. Impotent in the face of a massive death rate, especially among the young, doctors transfused patients with blood taken from survivors. A meta-analysis carried out in 2006 during the bird-flu pandemic looked at the use of convalescent therapy in 1918. Analysing eight studies involving 1,703 patients, the authors concluded that plasma therapy showed a significant reduction in the risk of mortality and that 'convalescent human H5N1 plasma could be an effective, timely, and widely available treatment that should be studied in clinical trials'.[4]

More recently, this approach has been used as a last resort during the SARS, MERS, Ebola and bird-flu epidemics. In 2003, there was a major outbreak of SARS in Taiwan. Three patients who had rapidly progressing disease with lung infiltration were treated with convalescent plasma. Their body temperatures dropped just one day after treatment, and viral-specific IgG and IgM antibodies were detected. X-rays showed improvement of lung inflammation and the virus was undetectable a day after transfusion. They survived, one female patient going on to deliver a baby some 13 months after being discharged. Interestingly, protective anti-SARS-CoV antibodies were detected in the newborn, suggesting passive transfer from the mother.[5] However, as there wasn't a control group, these remain anecdotal observations.

Patients who survive a critical Covid-19 infection appear to have the highest levels of anti-SARS-CoV-2 antibody. If the donor transfusions had indeed caused symptom resolution and recovery, then we can assume that the donor plasma contained antibodies that neutralized the virus. Such neutralizing antibodies are the holy grail of vaccinology, and protective IgG antibodies have been generated by the currently licensed vaccines. But will the vaccine-mediated secondary response prevent reinfection? Even though this has been shown in SARs-CoV-2-infected rhesus macaque monkeys, where infected animals could not be reinfected by the virus,[6] humans have, unfortunately, been shown to be reinfected following vaccination.

When should we administer the convalescent serum? Virus levels typically peak during the first week of infection, and the primary response that follows some seven to ten days later clears the virus in most patients infected with SARS-CoV-2.[7] Thus convalescent sera should be given early on in the infection. Perhaps this is one reason why large-scale trials failed to demonstrate clinical benefit.

Several strategies have been used in the past. These include use of convalescent whole blood, plasma or serum, pooled human immunoglobulin, high levels of human immunoglobulin and polyclonal or monoclonal antibodies. Such passive immunization methods have been used in a wide range of viral infections: influenza, chickenpox, measles, H1N1, H5N1 bird flu, Lassa fever, SARS and MERS. All publications on the use of convalescent sera in SARS show the same outcomes: viral concentrations decreased, and patients improved.[8]

SARS-CoV-2, the virus that has, at the time of writing, killed nearly 5 million people worldwide, looks beautiful. Depictions of its structure are constantly on our television screens. On its surface a crown or 'corona' of club-shaped sticks projects outwards. SARS-CoV-2 uses these 'clubs' to attach itself to cells of the upper respiratory tract and the lung. A protein called 'spike' found at the end of the club binds to ACE-2 receptors on the host cell. Antibodies binding to this spike protein can stop the virus in its tracks, blocking entry and effectively neutralizing it. Such neutralizing antibodies have been found in patients infected with the SARS, MERS and Covid-19 coronaviruses, and these antibodies found in serum are thought to be responsible for the amelioration of symptoms, allowing for recovery.[9]

*

There *is* a way of getting a limitless supply of ready-made neutralizing antibodies: Köhler and Milstein's monoclonal antibodies, as described in Chapter 26. Neutralizing monoclonal antibodies have been used in influenza, Ebola, HIV and rabies.[10] The antigen target for monoclonals raised against SARs-CoV-2 is the receptor-binding domain of the spike protein. Luckily, these proteins are highly immunogenic and don't appear to be very different from one patient to another.

Several groups have developed monoclonal antibodies against SARS-CoV-2. One such group made 206 monoclonal antibodies directed against the receptor-binding domain of the spike protein. The technique they used highlights how new technology is being used to produce human monoclonal antibodies, a far cry from the original approach used by Köhler and Milstein in the 1970s.

In this study, scientists isolated spike-protein-binding B cells from blood taken from eight SARS-CoV-2-infected patients. Using gene technology, they then isolated the spike antigen-binding IgG variable heavy and light chain genes and inserted these genes into plasmids. The plasmids, circular rings of DNA found in bacteria, already had the genes coding for the constant regions of human IgG heavy and light chains put into them. They were now armed with all the genes needed to make specific IgG monoclonal antibodies against the spike protein of SARS-CoV-2. The plasmids, capable of independent replication, were then put into cultured cells that secreted the monoclonal antibody translated from the antibody genes in the plasmids.

These human monoclonal antibodies can bind and neutralize live SARS-CoV-2, showing promise as future prophylactic and therapeutic agents in the battle against Covid-19.[11] At the time of writing, monoclonal antibodies have been prepared from B cells isolated from Covid-19 patients, using a range of techniques involving humanized mice, rabbits, monkeys and llamas.[12]

Monoclonal antibodies have also been approved and administered to Covid-19 patients, but their use is not widespread. Additionally, as the virus mutates and novel spike proteins are generated, the original monoclonal antibodies may not be able to recognize the altered spike protein and effectively block the new strains. In an interview on CNN, Dr Anthony Fauci, America's leading scientist coordinating the battle against Covid-19, cautioned about getting too optimistic about using antibodies against SARS-CoV-2. He highlighted the problem of antibody enhancement, a phenomenon seen in dengue fever and respiratory syncytial viral infection. Here antibodies bind to the virus and facilitate entry into cells, leading to increased replication and more severe disease. This phenomenon can allow viruses to enter immune cells that, while not having the virus-binding receptor, have receptors that bind the Fc or non-antigen-binding part of IgG. Therefore, a virus can use the antibody bound to the Fc receptor to enter the cell.[13]

We still don't know if this phenomenon happens in SARS-CoV-2 infection. At the time of writing this seems unlikely. But it has been observed in other coronavirus diseases, including SARS. In one study, passive transfer of anti-spike antibodies generated by vaccination worsened lung disease in macaques. Additionally, these antibodies failed to neutralize the virus when the animals were challenged with live virus. These antibodies also caused lung injury early in the infection by changing a beneficial macrophage response to one that caused inflammation. These 'inflammatory' macrophages secreted high levels of IL-6, which contributed to lung pathology. Thus, some macrophages are pro-inflammatory while others can have an inflammation-resolving role. Patients who succumbed to SARS were shown to have a dominance of pro-inflammatory macrophages in their lungs.[14]

We cannot rule out the possibility that the same thing will happen with SARS-CoV-2, and this risk must be considered when considering large-scale use of convalescent plasma and monoclonal antibodies against the virus. Another unknown is the speed at which SARS-CoV-2 can change its spike protein, thus stopping neutralizing antibodies from binding to the virus.

But could we use monoclonal antibodies to tackle the cytokine storm and stop people dying? In SARS and MERS, a suppressed early type I interferon response followed by an enhanced late cytokine response led to lung inflammation and death. A similar pattern has been seen in Covid-19. The suboptimal early interferon response allowed viral replication to continue unabated, causing an influx of neutrophils and macrophages into the lung, in turn resulting in unchecked inflammation. This heralds the cytokine storm which lead to ARDS, the leading cause of death in SARS, MERS and Covid-19. The pro-inflammatory storm troopers led by IL-6 can also cause the death of T cells, which further allows the virus to get the upper hand. Increased IL-6 levels, along with low lymphocyte counts, correlate with disease severity in MERS, SARS and Covid-19.[15]

Monoclonal antibodies targeting interleukin 6 or the IL-6 receptor are now included in our armamentarium against Covid-19. These drugs are used to reduce inflammation in diseases like rheumatoid arthritis. The anti-IL-6 monoclonal drug tocilizumab was also approved in 2017 for treating cytokine release syndrome.

Chapter 31

THE RACE

Pittsburgh, Pennsylvania, 16 May 1953. A black-and-white photograph captures a moment in history. A smiling nine-year-old is sitting down, his left arm held by a man in a white coat who is holding a syringe in his right hand. A nurse stands behind him, and his mother, sitting to his right, is also smiling, nervously.

'I just hated injections,' recalls Peter Salk, now 76. 'So, my father came home with polio vaccine and some syringes and needles that he sterilized on the kitchen stove, boiling them in water, and lined us kids up and then administered the vaccine.'[1]

In 1947, Jonas Salk, the son of Polish immigrants, was given the task of coming up with a vaccine against polio. Americans were terrified of the disease, with good reason. In 1952 alone, nearly 58,000 people in the United States had contracted the condition. Some 21,000 had a degree of paralysis and the virus had claimed over 3,000 lives.[2]

The infection was particularly severe in children, invading the central nervous system and in some cases causing paralysis and death. In its most destructive phase, polio paralysed the diaphragm and chest-wall muscles, forcing patients to be imprisoned in iron lungs for weeks, sometimes years.

Summertime was dreaded. It was called the 'polio season', when public swimming pools were closed down and people were asked not to sit close to each other inside cinemas and theatres. They were also advised to wash their hands frequently.[3]

For the next seven years, Salk, who had developed a vaccine against influenza, worked '16-hour days, seven days a week', finally coming up with a vaccine that was safe to use.[4] The vaccine was made by 'killing' the virus with formaldehyde: it couldn't infect cells and replicate.

After successful tests on monkeys, Salk was ready to test his vaccine on children. On 2 July 1952, he injected 30 children living at the D.T. Watson Home for Crippled Children in Pittsburgh. These children were not at risk of contracting polio: they were already severely disabled by the disease. Salk found that vaccination increased the levels of polio-specific antibodies in the children. There were no side effects. A few weeks later, he injected another group of children – who had never contracted polio – at an institution then called the Polk State School for the 'retarded and feeble-minded', with his killed polio vaccine.[5]

In 1954, he tested the vaccine on over 2 million children, known as the 'polio pioneers'. The vaccine was deemed safe on 12 April 1955, and the United States began mass vaccinations. The country was free of polio by 1979.[6]

*

In early 2020, scientists from Sinovac Biotech in Beijing used a similar approach to produce a vaccine against SARS-CoV-2. They gave a group of macaques multiple doses of a whole inactivated SARS-CoV-2 vaccine made from a virus isolated from lung fluid aspirated from a Covid-19 patient. After growing in tissue culture, the viruses were harvested, inactivated by beta-propiolactone and mixed with alum adjuvant, a compound that makes the vaccine cause a stronger immune response.

The vaccine protected the macaques from developing pneumonia when challenged by live virus, and crucially did not show antibody-dependent enhancement. When compared to control animals who did not receive the vaccine, the levels of pro-inflammatory cytokines such as IL-6 were not elevated in the vaccinated animals. Additionally, post-mortem examination showed no lung pathology.[7]

Because they are relatively easy to make, inactivated viral vaccines will give the fastest results when tested during a pandemic. Inactivated vaccines first appeared at the end of the nineteenth century. Early efforts by Roux and Chamberland at the Pasteur Institute led to such vaccines being made against cholera, typhoid and plague. As we saw in Chapter 13, Almroth Wright used an inactivated typhoid vaccine on thousands of soldiers.[8] The first influenza vaccines were inactivated viruses.

The first efforts at immunizations used a different approach: weakened pathogens. Pasteur used attenuated pathogens as vaccines against anthrax and rabies. Attenuation was achieved by passage in animals, heat or exposure to oxygen. Pasteur used an attenuated vaccine at Pouilly-le-Fort during the first controlled clinical trial of a vaccine in history. The BCG vaccine strain was attenuated by passaging the bovine tubercle 230 times in an artificial medium.[9] The arrival of techniques enabling tissue culture led to the development of live vaccines against polio, measles, mumps, rubella and chickenpox. Growing viruses in tissue culture led to a loss in pathogenicity, but the spectre of reversion always remained a possibility.

*

In 1955, the Salk vaccine hit a roadblock. A defective batch caused around 40,000 cases of polio. Nearly 200 children became paralysed and ten died. The reason was inefficient killing of the virus. The defective vaccine batch was traced back to the Cutter Laboratories, a family-run company in Berkeley, California. Inactivated vaccines were not risk-free.[10]

At the same time, Albert Sabin developed an attenuated live polio vaccine that could be administered orally. This trivalent vaccine, made by passage in animals and tissue culture, was tested in Mexico and the Soviet Union and eventually replaced the Salk vaccine in 1962.[11] However, there have been rare cases of paralytic polio as a result of reversion. In the UK, the oral polio vaccine has now been replaced by a formaldehyde-inactivated vaccine.[i]

But what if we could remove the SARS-CoV-2 genes that cause pathogenicity, and put the harmless spike-protein gene into a vector virus? As living vaccines are better at generating T- and B-cell responses, this could be an ideal vaccine strategy against Covid-19. The first vaccine group to get off the blocks used this approach.

*

Scientists from the Jenner Institute, Oxford, England, carried out initial work in April 2020. A chimpanzee adenovirus vector carrying the SARS-CoV-2 spike-protein gene was injected into six rhesus macaques. Three monkeys – the control group – were injected with an adenovirus virus lacking the gene. These non-human primates can be infected with SARS-CoV-2 and the virus can spread to the lungs, mimicking the human disease. Twenty-eight days later, when challenged with a high dose of live SARS-CoV-2, the vaccinated animals showed less virus in the lungs and upper respiratory tract compared to control animals. Spike-specific neutralizing antibodies were found 14 days after vaccination in the vaccinated group but not in the control group. SARS-CoV-2 spike-specific T cells also appeared in the vaccinated macaques. None of the vaccinated animals developed pneumonia. In contrast to the SARS study described in Chapter 30, there was no evidence of immune-enhanced disease. Seven days after inoculation all the monkeys were euthanized. There was no evidence of virus or lung pathology in any of the vaccinated animals. In contrast, lung tissue from two out of three control animals showed signs of viral pneumonia: alveoli invaded by lymphocytes and macrophages. Interestingly, following challenge with live virus the vaccinated monkeys still shed virus from the nose, but the disease did not spread to the lungs.[12] This group had previously shown that rhesus macaques inoculated with a vaccine containing the spike protein of the MERS virus were protected from a MERS-CoV challenge.[13]

Human trials led to the emergence of the Oxford AstraZeneca vaccine, which was approved in several countries. At the time of writing the vaccine, supported by real-world data, has been shown to be highly effective in preventing severe disease, hospitalization and death.

Another approach involves introducing the genetic material coding for the spike protein directly into a person. The DNA or messenger RNA (mRNA) will then translate the genetic information and produce the viral protein which will provoke an immune response. DNA vaccines have many advantages. Large amounts can be produced quickly, and the vaccines are temperature-stable.

One group has produced a DNA vaccine containing the SARS-CoV-2 spike-protein gene. This group had previously constructed a DNA-based vaccine for MERS, which had protected vaccinated macaques when challenged with the virus.[14] The Covid-19 vaccine containing a similar replicating plasmid produced neutralizing antibodies and T cells when injected into mice and guinea pigs.[15]

But the biggest breakthrough involved the development of vaccines that used mRNA constructs as vaccines. The Pfizer-BioNTech and Moderna vaccines, both RNA vaccines, have also been approved worldwide and millions have been vaccinated. As with the Oxford AstraZeneca vaccine, these mRNA vaccines have been shown to be highly effective in preventing serious disease. In the Moderna, the mRNA transcribing the spike protein of SARS-CoV-2 is encapsulated in a lipid nanoparticle and injected into the upper arm muscle.

*

Antibodies produced in immune tissue lying underneath the respiratory epithelium can prevent viruses binding onto epithelial cells.

Secretory IgA (sIgA) antibodies found in mucosal secretions, such as tears, saliva and breast milk, demonstrate antiviral functions, but the duration of the response is much shorter when compared to serum-antibody responses.

Mucosal vaccines exploit the existence of the common mucosal immune system where immune responses elicited against an antigen at one site generate a response at an anatomically distant site. An ingested oral vaccine will generate sIgA responses in both the gut and the respiratory mucosa. Sabin's oral polio worked in this way: the attenuated virus multiplied in the gut and immune responses elicited in the gut mucosal tissue protected the airways.

This approach has been tried against viral infections. In 2015, Vaxart carried out a randomized, double-blind, placebo-controlled trial using a recombinant, non-replicating adenovirus vaccine expressing the haemagglutinin gene of the influenza A (H1N1) virus. Compared to controls, a single dose of this oral vaccine elicited neutralizing antibodies against influenza virus in the vaccinated subjects.[16]

Stabilitech, a UK-based company, has developed a non-replicating adenovirus vector vaccine containing DNA coding for SARS-CoV-2 spike antigen. Trials involving these oral vaccines are underway.

Which vaccine will work against the emerging variants? This is the challenge, as the virus will evolve to escape the vaccines. The world waits for the miracle. Who will come up with a vaccine that is effective against these new challenges? The winners will be our saviours, the world will be grateful, and there will be huge rewards.

*

Jonas Salk died in 1995. In 1952, a journalist asked him who would own the patent for his polio vaccine. The doctor smiled and replied: 'The people, I would say. There is no patent. Could you patent the sun?'[17]

i   There are certain considerations that need to be taken into account when a vaccine is designed. The incubation time of the pathogen is important, as a very short incubation time would not allow for the generation of a T-cell response; therefore, a vaccine generating blocking antibodies may be more desirable.

Activation of the appropriate branch of the immune system is important as different pathogens or their products are neutralized by diverse types of response. For example, bacterial toxins require antibodies that prevent them binding to their receptors on host cells, whereas viral infections and intracellular bacterial infections require T-cell responses that will destroy the infected cell. To generate T-cell responses, the vaccine should ideally be able to replicate within host cells. So activation of the appropriate branch of the immune system is a crucial prerequisite. To design a vaccine against a particular pathogen one must know the immune correlates of protection: that is, which type of response can destroy the pathogen. Is it neutralization antibodies, or cytotoxic T cells, or cytokines? Childhood immunizations require multiple boosters: the persistence of maternal antibodies can block epitopes, as in the diphtheria, tetanus and pertussis (DPT) vaccine.

The current routine immunization of children includes the use of six vaccines given as a single injection. This vaccine (diphtheria, tetanus, acellular pertussis, inactivated polio, *Haemophilus influenzae* type b and hepatitis B) is given at eight weeks. So what are the constituents of this complex vaccine? The *Haemophilus influenzae* type b vaccine is a capsular polysaccharide derived from *Haemophilus influenzae*. As the polysaccharide only activates B cells producing a predominantly IgM response with little or no T helper or memory cell activation, the vaccine is covalently linked to tetanus toxoid. This increases its immunogenicity, activating T helper cells and also enabling class-switching to occur. Diphtheria and tetanus vaccines are toxoids that will generate antitoxin antibodies. The whooping-cough vaccine is an acellular vaccine made from highly purified macromolecules of *Bordetella pertussis*. To improve immunogenicity, these components are treated with formaldehyde or glutaraldehyde and then adsorbed onto adjuvants: either aluminium phosphate or aluminium hydroxide. Hepatitis B vaccine is an inactivated vaccine.

Chapter 32

LIFE, DEATH AND THE IMMUNE SYSTEM

But the worldwide mortality statistics are grim. Globally, almost 5 million people have died after contracting Sars-Cov-2. In the United States alone the death toll is over 700,000. The virus that jumped species from bats to man is still among us – the pandemic is not over.

During the peak in 2020, I tried to find out how many people in my age group had died of Covid-19. Out of the 35,000 people who had succumbed in the UK, 5 per cent were in their fifties. This jumped to 11 per cent for people in their sixties, 23 per cent for people in their seventies and 59 per cent in those over 80. Of all deaths relating to Covid-19, 88 per cent had occurred among people over 65.[1] This pattern has persisted.

The data reveals the pattern seen throughout the world: the chance of dying from Covid-19 increases with age. In contrast, morbidity and mortality in young children is much lower. The reasons for this are unknown.

Age is the most important risk factor for morbidity and mortality from Covid-19. There could be several reasons for this, but an important one involves the decline of the immune system with ageing.

As we get older, our immune system, like all other systems, goes into a downturn. The thymus gland, the primary lymphoid organ where T lymphocytes acquire their receptors, enabling them to engage pathogens and coordinate immune responses, atrophies with age. The weight of the gland peaks at puberty and then declines as we get older. The thymus gland is involuted – completely replaced by fatty tissue – by the age of 60 in men and by 70 in women. Obesity and smoking are associated with accelerated thymic involution.[2] Incidentally, obesity also increases the risk of infection.[3]

A consequence of thymic involution is a decline in circulating naive T cells, making it difficult to generate effective immune responses to new pathogens and vaccines.

Immunosenescence is the term used to describe the decline in immunity due to ageing. Both innate and adaptive immune systems undergo this process. As T cells are vital in the clearance of viral infections like SARS-CoV-2, let's look at what happens to T cells as we age.

Firstly, in addition to having fewer naive cells, the T cells we have function less well. Many terms describe these aged T cells: 'senescent', 'exhausted', 'inflammatory'.[4] The bottom line is that these T cells mount less effective immune responses over time. There are many reasons why ageing T cells perform less than optimally, but these mechanisms have not been fully elucidated.

One reason involves telomere shortening. 'Telomeres' are caps found at the end of chromosomes, and, like the plastic or metal caps on the ends of shoelaces that protect them from fraying, protect the chromosome ends, which get shorter and unstable during cell division. Every time a cell divides the telomeres get shorter. Shortened telomeres can therefore be looked upon as markers of immune-cell senescence as these cells are more likely to undergo apoptosis, cell suicide. Having short telomeres is associated with greater morbidity and mortality in several age-related diseases. The enzyme telomerase protects telomere length and helps mitigate cellular ageing. Having longer telomeres and higher telomerase activity can therefore indicate an effective immune system.

What causes telomere shortening? The culprit could be inflammation. Older people have low-grade inflammation characterized by higher levels of the pro-inflammatory cytokines like IL-6 and lower levels of the anti-inflammatory ones like IL-10. This process, called 'inflamm-ageing', is thought to be involved in the pathogenesis of age-related diseases like atherosclerosis, diabetes and Alzheimer's disease.

This low level of inflammation could explain why telomeres in older people are shorter and the levels of telomerase lower, although the relationship is thought to be bidirectional.[5] An inverse relationship has been seen between increasing age and shorter telomeres in T cells, especially in those aged 60 and over, and low telomerase activity has been found to be associated with T-cell senescence.[6]

It's not just T-cell function that declines with age. The whole immune system progressively deteriorates. There are fewer circulating monocytes and dendritic cells; phagocytic function of neutrophils declines, and there is a gradual diminution of the T- and B-cell repertoire, leading to reduced antibody and cell-mediated responses.[7] These changes could explain why the elderly are more vulnerable to infections like Covid-19. Their immune systems find it difficult to mount effective immune responses to new antigens, and SARS-CoV-2 is a novel pathogen that humanity has never encountered before. The elderly might also find it difficult to mount responses against a vaccine. Studies have shown that the elderly respond poorly to the influenza vaccine.[8]

At the time of writing, the current Covid-19 vaccines appear to be effective in the older age groups. But it is early days yet, and whether the protective immunity generated in these individuals is persistent over the coming months remains to be seen.

Are there interventions that might assist the immune system in its battle against the virus? Is there anything we as individuals can do to optimize our immune defences?

*

We can exercise. Currently no studies have specifically looked at the effect of exercise on coronavirus infection, but we can draw some conclusions from the literature. Studies have consistently shown that moderate to vigorous exercise can improve immune responses to vaccination. Exercise can also damp down chronic low-grade inflammation and reduce obesity, a known risk factor for morbidity and mortality in Covid-19. Exercise also improves glycaemic control in diabetes, another risk factor in Covid-19. Several studies have shown that exercise can ameliorate morbidity in a wide variety of viral infections: influenza, rhinovirus, herpesvirus, Epstein–Barr virus and chickenpox.[9] Studies in mice have shown that moderate-intensity exercise during an active influenza infection can reduce mortality by countering adverse cellular and cytokine activity in the lungs.[10]

In the late 1960s, it was recognized that the immune system interacted with the nervous and endocrine systems and consequently psychological stress and behaviour impacted on immune parameters. This was the birth of a new field: psychoneuroimmunology. The brain communicated with the immune system through the autonomic nervous system and hormones. Surprisingly, cells of the immune system were found to have receptors for neurotransmitters, and cytokines secreted by immune cells affected the brain and mind. Thus, there are bidirectional pathways interlinking the brain and the immune system. Studies found that certain behaviours impacted on the immune system. Pavlovian conditioning could suppress or enhance immune responses, and stressful life experiences and negative mood states like depression were detrimental to immune function. The immune system appears to be exquisitely sensitive to mental states and behaviour. Unsurprisingly, immune-mediated diseases were particularly responsive.[11]

This makes manipulation possible. One example of a behavioural intervention is mindfulness. A meta-analysis carried out in 2016 scrutinized all the eligible randomized controlled studies that

looked at the effects of mindfulness on immunity.[12] They identified three areas where there was clear proof of benefit: reductions
in circulating levels of inflammatory markers, increases in T-cell
counts and increases in telomerase activity. Their analysis also
points to promising areas of future investigation.

In one study, levels of the pro-inflammatory cytokine IL-6
decreased in breast-cancer patients following participation in a
six-week mindfulness course compared to non-participants in a
control group.[13] Another study showed an increase in the anti-
inflammatory protein IL-10 in ulcerative-colitis patients following
an eight-week meditation course. Levels of IL-10 in the control
group participating in a matched mind–body medicine programme
did not change.[14]

Another protein studied is the ubiquitous transcription factor
NF-κB. This protein is activated when viral RNA inside a host cell
is sensed by internal receptors. When activated, NF-κB induces
an expression of pro-inflammatory cytokines. Two studies demonstrated a reduction in the expression of NF-κB after subjects
(elderly adults experiencing chronic loneliness[15] and breast cancer
patients[16]) participated in an eight-week mindfulness course. A
waiting-list control group showed no such reduction. In contrast,
NF-κB levels increase following psychological stress, and this could
explain how stress contributes to diseases linked to persistent
inflammation.[17]

As we saw in Chapter 28, a decline in $CD4^+$ T-cell count heralds
disease progression and adverse outcomes in HIV patients. Low
lymphocyte counts have been consistently observed in Covid-19
patients who develop severe disease. Several randomized controlled
studies showed an increase or a buffered decline in blood T-cell
counts and/or the percentage of activated T cells after mindfulness
meditation in people diagnosed with HIV or breast cancer.[18]

In all three randomized controlled studies that measured telomerase activity, mindfulness was shown to increase the levels of the

enzyme after subjects followed a mindfulness programme. The studies included a diverse group of subjects: meditation retreat participants, overweight/obese women and breast-cancer patients. The effects were shown to be dose-dependent, where adherence to the practice reflected telomerase increases.[19]

How mindfulness, a purely mental exercise, changes the immune system is not known, but Elizabeth Blackburn, the Australian scientist who won the 2009 Nobel Prize for the discovery of telomeres and telomerase, thinks it's partly down to telomerase. 'Nobody had any idea,' she said in a recent interview, 'that meditation and the like, which people can use to reduce stress and increase wellbeing, would be having their salutary and well-documented useful effects in part through telomeres.'[20]

*

In December 1851, Pasteur wrote, 'I am at the edge of mysteries, and the veil is getting thinner and thinner.' Covid-19 has demonstrated that we are still at the edge; and a lot more remains to be discovered.

Acknowledgements

I would like to thank Professor Neville Punchard for his invaluable advice and guidance. I am indebted to Dr Steve Anderson for the illustrations. A very special thanks to Mr Ian Howe, and to Mr Andrew Lownie, who encouraged me to write section three of the book. I would also like to thank my colleagues Dr Iain Nicholl, Dr James Vickers, Dr Kathryn Dudley, Dr Paraskevi Goggolidou, Dr Kesley Attridge and Professor Paul Kirkham for their advice. I would also like to thank Christian Müller from Hero Press.

Endnotes

INTRODUCTION

1   Zhe Xu et al., 'Pathological findings of COVID-19 associated with acute respiratory distress syndrome', *Lancet Respiratory Medicine* 2020; 8(4): 420–2.

2   Rudragouda Channappanavar and Stanley Perlman, 'Pathogenic human coronavirus infections: causes and consequences of cytokine storm and immunopathology', *Seminars in Immunopathology* 2017; 39(5): 529–39.

3   David M. Morens and Jeffery K. Taubenberger, 'Influenza cataclysm 1918', *New England Journal of Medicine* 2018; 379: 2285–7.

4   V. C. Vaughan, *A Doctor's Memories* (Indianapolis: Bobbs-Merrill, 1926), 383–4.

5   Jeffery K. Taubenberger et al., 'Initial genetic characterization of the 1918 "Spanish" influenza virus', *Science* 1997; 275(5307): 1793–6.

6   Ann H. Reid, Thomas G. Fanning, Johan V. Hultin and Jeffery K. Taubenberger, 'Origin and evolution of the 1918 "Spanish" influenza virus hemagglutinin gene', *Proceedings of the National Academy of Science of the USA* 1999; 96(4): 1651–6.

7   Terrence M. Tumpey et al., 'Characterization of the reconstructed 1918 Spanish influenza pandemic virus', *Science* 2005; 310(5745): 77–80.

8   Kirsty R. Short, Katherine Kedzierska and Carolien E. van de Sandt, 'Back to the future: lessons learned from the 1918 influenza pandemic front', *Frontiers in Cellular and Infection Microbiology* 2018; 8: 1–19.

9   Darwyn Kobasa et al., 'Aberrant innate immune response in lethal infection of macaques with the 1918 influenza virus', *Nature* 2007; 445(7125): 319–23.

10 Irani Thevarajan et al., 'Breadth of concomitant immune responses prior to patient recovery: a case report of non-severe COVID-19', *Nature Medicine* 2020; 26(4): 453–5.

11 Michael J. Ciancanelli et al., 'Life-threatening influenza and impaired interferon amplification in human IRF7 deficiency', *Science* 2015; 348(6233): 448–53.

12 Kobasa et al., 'Aberrant innate immune response'.

13 Yvan Jamilloux et al., 'Should we stimulate or suppress immune responses in COVID-19? Cytokine and anti-cytokine interventions', *Autoimmunity Reviews* 2020; 19(7): 102567.

14 Sarah Shalhoub, 'Interferon beta-1b for Covid-19', *Lancet* 2020; 395(10238): 1670–1.

15 Moritz Anft et al., 'COVID-19 progression is potentially driven by T-cell immunopathogenesis', medRxiv preprint, 2020. doi: https://doi.org/10.1101/2020.04.28.20083089.

16 Fei Zhou et al., 'Clinical course and risk factors for mortality of adult inpatients with COVID-19 in Wuhan, China: a retrospective cohort study', *Lancet* 2020; 395(10229): 1054–62.

17 Kobasa et al., 'Aberrant innate immune response'.

CHAPTER 1

1 *The Landmark Thucydides*, ed. Robert B. Strassler (New York: Simon & Schuster, 2008), 120.

2 Abú Becr Mohammed ibn Zacaríyá ar-Rází (commonly called Rhazes), *A Treatise on the Small-Pox and Measles*, tr. W. A. Greenhill (London: Sydenham Society, 1848), 34.

3 Abbas M. Behbehani, 'The smallpox story: life and death of an old disease', *Microbiological Reviews* 1983; 47(4): 455–509.

4 Stefan Riedel, 'Edward Jenner and the history of smallpox and vaccination', *Baylor University Medical Center Proceedings* 2005; 18(1): 21–5.

CHAPTER 2

1 Lady Mary Wortley Montagu, *Letters of the Right Honourable Lady Montagu: Written during Her Travels in Europe, Asia and Africa* (Aix: Anthony Henricy, 1796), i, 167–9; letter 36, to Mrs. S. C. from Adrianople, n.d. Available at https://sourcebooks.fordham.edu/mod/montagu-smallpox.asp. See also Lady Mary Wortley-Montagu, *The Turkish Embassy Letters*, ed. Malcolm Jack (London: Virago, 1994).

2 Abbas M. Behbehani, 'The smallpox story: life and death of an old disease', *Microbiological Reviews* 1983; 47(4): 455–509.

CHAPTER 3

1 Henry J. Parish, *Victory with Vaccines: The Story of Immunization* (Edinburgh: E. & S. Livingstone, 1968), 16.

2 Arthur M. Silverstein, *A History of Immunology* (2nd edn, Amsterdam: Academic Press, 2009), 291.

3 Quoted in Parish, *Victory with Vaccines*, 16.

4 Edward Jenner, *An Inquiry into the Causes and Effects of the Variolae Vaccinae* (London: printed for the author, 1798).

5   Quoted in Abbas M. Behbehani, 'The smallpox story: life and death of an old disease', *Microbiological Reviews* 1983; 47(4): 455–509.
6   Frank Fenner et al., *Smallpox and Its Eradication* (Geneva: WHO, 1988), 175.

CHAPTER 4

1   Antony van Leewenhoeck, 'Observations, communicated to the publisher', *Philosophical Transactions of the Royal Society of London* 1677; 12(133): 821–31.
2   Quoted in René Vallery-Radot, *The Life of Pasteur*, tr. R. L. Devonshire (London: Constable, 1906), 108.
3   Alan Gillen and Frank J. Sherwin III, 'Louis Pasteur's views on creation, evolution, and the genesis of germs', *Answers Research Journal* 2008; 1: 43–52.
4   Patrice Debré, *Louis Pasteur*, tr. Elborg Forster (Baltimore: Johns Hopkins University Press, 2000), 156.
5   Ibid., 124.
6   Ibid., 126.

CHAPTER 5

1   S. M. Blevins and M. S. Bronze, 'Robert Koch and the "golden age" of bacteriology', *International Journal of Infectious Diseases* 2010; 14(9): 744–51.
2   Ibid.
3   Ibid.

CHAPTER 6

1   M. Harper, J. D. Boyce and B. Adler, 'Pasteurella multocida pathogenesis: 125 years after Pasteur', *FEMS Microbiology Letters* 2006; 265(1): 1–10.
2   Henry J. Parish, *Victory with Vaccines: The Story of Immunization* (Edinburgh: E. & S. Livingstone, 1968), 23.
3   Ibid, 24.
4   Louis Pasteur, 'Summary report of the experiments conducted at Pouilly-le-Fort near Melun, on the anthrax vaccination, 1881', *Yale Journal of Biology and Medicine* 2002; 75(1): 59–62.
5   Ibid.
6   Quoted in Agnes Ullmann, 'Pasteur–Koch: distinctive ways of thinking about infectious diseases', *Microbe* 2007; 2(8): 383–7.
7   R. Koch, 'On the anthrax inoculation' [1882], in *Essays of Robert Koch*, tr. K. C. Carter (New York: Greenwood Press, 1987), 100.
8   Ibid, 32.

CHAPTER 7

1   Henry J. Parish, *Victory with Vaccines: The Story of Immunization* (Edinburgh: E. & S. Livingstone, 1968), 26.
2   Quoted in Anita Guerrini, *Experimenting with Humans and Animals: From Galen to Animal Rights* (Baltimore and London: Johns Hopkins University Press, 2003), 101.

CHAPTER 8

1   S. M. Blevins and M. S. Bronze, 'Robert Koch and the "golden age" of bacteriology', *International Journal of Infectious Diseases* 2010; 14(9): 744–51.
2   Hippocrates, 'Book 1 – of the epidemics', in *The Genuine Works of Hippocrates*, tr. Francis Adams (London: Sydenham Society, 1849). See John Frith, 'History of tuberculosis. Part 1 – phthisis, consumption and the white plague', *Journal of Military and Veterans' Health* 2014; 22(2): 29–35.
3   WHO tuberculosis factsheet N104, October 2015.
4   S. K. Jindal, ed., *Textbook of Pulmonary and Critical Care Medicine* (New Delhi: Jaypee Brothers, 2011), 549.
5   See John Frith, 'History of tuberculosis. Part 1 – phthisis, consumption and the white plague', *Journal of Military and Veterans' Health* 2014; 22(2): 29–35.
6   Hervey Allen, *Israfel: The Life and Times of Edgar Allan Poe* (New York: George H. Doran, 1926), ii, 519.

CHAPTER 9

1   Quoted in Olga Metchnikoff, *Life of Élie Metchnikoff 1845–1916* (Boston and New York: Houghton Mifflin, 1921), 116–17.
2   Quoted in Stefan H. E. Kaufmann, 'Immunology's foundation: the 100-year anniversary of the Nobel Prize to Paul Ehrlich and Elie Metchnikoff', *Nature Immunology* 2008; 9(7): 705–12.
3   Quoted ibid.
4   D. Neil Granger and Elena Senchenkova, *Inflammation and the Microcirculation* (San Rafael, CA: Morgan & Claypool Life Sciences, 2010), 5.
5   Ibid.
6   Ibid.

CHAPTER 10

1   Henry J. Parish, *Victory with Vaccines: The Story of Immunization* (Edinburgh: E. & S. Livingstone, 1968), 45.

2 E. Roux and A. Yersin, 'Contribution à l'étude de la diphthérie', *Annales de l'Institut Pasteur* 1888; 2(12): 629–61. See also Jack H. Botting, 'Diphtheria: understanding, treatment and prevention', in Jack H. Botting, *Animals and Medicine: The Contribution of Animal Experiments to the Control of Disease* (Cambridge: Open Book Publishers, 2015), 63–5.

3 E. Behring, 'Untersuchungen ueber das Zustandekommen der Diphtherie-Immunität bei Thieren', *Deutsche Medizinische Wochenschrift* 1890; 50: 1145–8. See also S. H. Kaufmann, 'Remembering Emil von Behring: from tetanus treatment to antibody cooperation with phagocytes', *mBio* 2017; 8(1): e00117–17.

4 Inaya Hajj Hussein et al., 'Vaccines through centuries: major cornerstones of global health', *Frontiers in Public Health* 2015; 3(1): 269.

5 Quoted in Parish, *Victory with Vaccines*, 48–9.

6 Quoted ibid, 49.

7 Ibid, 47.

8 Ibid, 66.

9 Ibid, 54.

10 Niels K. Jerne, 'The generative grammar of the immune system', Nobel Lecture, 8 December 1984. Available at https://www.nobelprize.org/prizes/medicine/1984/jerne/lecture/.

CHAPTER 11

1 Quoted in Stefan H. E. Kaufmann, 'Immunology's foundation: the 100-year anniversary of the Nobel Prize to Paul Ehrlich and Elie Metchnikoff', *Nature Immunology* 2008; 9(7): 705–12.

2 Paul Ehrlich, 'Partial cell functions', Nobel Lecture, 11 December 1908. Available at https://www.nobelprize.org/uploads/2018/06/ehrlich-lecture.pdf.

3 Ibid.

4 Ibid.

5 Ian R. Mackay, 'Travels and travails of autoimmunity: a historical journey from discovery to rediscovery', *Autoimmunity Reviews* 2010; 9(5): 251–8.

CHAPTER 12

1 Richard Pfeiffer's journal, quoted in Alfred I. Tauber and Leon Chernyak, *Metchnikoff and the Origins of Immunology: From Metaphor to Theory* (Oxford: Oxford University Press, 1991), 159.

2 Ibid.

CHAPTER 13

1   Jonathan Dworkin and Siang Yong Tan, 'Jules Bordet (1870–1961): pioneer of immunology', *Singapore Medical Journal* 2013; 54(9): 475–6.
2   Ibid.
3   John L. Turk, 'Almroth Wright: phagocytosis and opsonization', *Journal of the Royal Society of Medicine* 1994; 87(10): 576–7.
4   Ibid.
5   Ibid.
6   Ibid., quoting George Bernard Shaw, *The Doctor's Dilemma* (1906), Act I.

CHAPTER 14

1   Gunver S. Kienle, 'Fever in cancer treatment: Coley's therapy and epidemiologic observations', *Global Advances in Health and Medicine* 2012; 1(1): 92–100.
2   Ibid.
3   Ibid.
4   Ibid.
5   Edward F. McCarthy, 'The toxins of William B. Coley and the treatment of bone and soft-tissue sarcomas', *Iowa Orthopaedic Journal* 2006; 26: 154–8.
6   Matthew Tontonoz, 'What ever happened to Coley's toxins?', Cancer Research Institute [website] (2 April 2015). Available at https://www.cancerresearch.org/blog/april-2015/what-ever-happened-to-coleys-toxins.
7   H. E. Stephenson et al., 'Host immunity and spontaneous regression of cancer evaluated by computerized data reduction study', *Surgery, Gynecology and Obstetrics* 1971; 133: 649–55.
8   Tontonoz, 'What ever happened to Coley's toxins?'. The reduction in tumour burden correlated with an elevation of the levels of some cytokines.
9   Christopher R. Parish, 'Cancer immunotherapy: the past, the present and the future', *Immunology and Cell Biology* 2003; 81(2): 106–13.
10 Ibid.

CHAPTER 15

1   Hans P. Schwarz and Friedrich Dorner, 'Karl Landsteiner and his major contributions to haematology', *British Journal of Haematology* 2003; 121(4): 556–65.
2   Quoted ibid.
3   Karl Landsteiner and Merrill W. Chase, 'Experiments on transfer of cutaneous sensitivity to simple compounds', *Proceedings of the Society for Experimental Biology and Medicine* 1942; 49(4): 688–90.

4   Merrill W. Chase, 'Cellular transfer of cutaneous hypersensitivity to tuber-
    culin', *Proceedings of the Society for Experimental Biology and Medicine*
    1945; 59: 134–5.

CHAPTER 16

1   Marshall A. Lichtman et al., eds, *Hematology: Landmark Papers of the
    Twentieth Century* (San Diego: Academic Press, 2000), 24.
2   Ibid, 21.
3   Ibid.
4   Ian R. Mackay, 'Travels and travails of autoimmunity: a historical journey
    from discovery to rediscovery', *Autoimmunity Reviews* 2010; 9(5): 251–8.
5   Ibid.
6   Ibid.
7   Ibid.
8   Ibid.
9   Ibid.
10 Ibid.

CHAPTER 17

1   Murray Dworetzky et al., 'Portier, Richet and the discovery of anaphylaxis: a
    centennial', *Journal of Allergy and Clinical Immunology* 2002; 110(2): 331–6.
2   Ibid.
3   Quoted in William R. Clark, *At War Within: The Double-Edged Sword of
    Immunity* (New York and Oxford: Oxford University Press, 1995), 83.
4   Quoted in Dworetzky et al., 'Portier, Richet and the discovery of anaphylaxis'.
5   Juan M. Igea, 'The history of the idea of allergy', *Allergy* 2013; 68(8): 966–73.
6   Ibid.
7   C. von Pirquet, *Serum Sickness*, tr. B. Chick (Baltimore: Williams & Wilkins,
    1951), 119–20.
8   A. M. Silverstein, 'Clemens Freiherr von Pirquet: explaining immune complex
    disease in 1906', *Nature Immunology* 2000; 1(6): 453–5.
9   J. Ring and J. Gutermuth, '100 years of hyposensitization: history of allergen-
    specific immunotherapy (ASIT)', *Allergy* 2011; 66(6): 713–24.
10 Ibid.
11 Leonard Noon, 'Prophylactic inoculation against hay fever', *Lancet* 1911;
    177(4580): 1572–3.
12 John Freeman, 'Further observations on the treatment of hay fever by hypo-
    dermic inoculation of pollen vaccine', *Lancet* 1911; 178(4594): 814–7.
13 Ring and Gutermuth, '100 years of hyposensitization'.
14 Ibid.

15 Mary Hewitt Loveless and John R. Cann, 'Distribution of allergic and "blocking" activity in human serum proteins fractionated by electrophoresis convection', *Science* 1953; 117(3031): 105–8.

16 U. Blank, F. H. Falcone and G. Nilsson, 'The history of mast cell and basophil research – some lessons learnt from the last century', *Allergy* 2013; 68(9): 1093–101.

17 Ibid.

18 Henry H. Dale, 'The anaphylactic reaction of plain muscle in the guinea pig', *Journal of Pharmacology and Experimental Therapeutics* 1913; 4(3): 167–223.

19 Philip G. H. Gell and Robin R. A. Coombs, 'The classification of allergic reactions underlying disease', in Philip G. H. Gell and Robin R. A. Coombs, eds, *Clinical Aspects of Immunology* (Oxford: Blackwell, 1963).

20 Kimishige Ishizaka, Teruko Ishizaka and Margaret M. Hornbrook, 'Physiochemical properties of reaginic antibody', *Journal of Immunology* 1966; 97(6): 840–53.

21 A. P. Weetman, 'Graves' disease 1835–2002', *Hormone Research in Paediatrics* 2003; 59 (supp. 1): 114–18.

CHAPTER 18

1 Peter B. Medawar, *Memoir of a Thinking Radish: An Autobiography* (Oxford: Oxford University Press, 1986), 77.

2 Ibid, 80.

3 Thomas Gibson and Peter B. Medawar, 'The fate of skin homografts in man', *Journal of Anatomy* 1943; 77(4): 299–310.

4 Medawar, *Memoir of a Thinking Radish*, 83.

5 Leslie Brent, *A History of Transplantation Immunology* (San Diego: Academic Press, 1997), 71.

6 Medawar, *Memoir of a Thinking Radish*, 111.

7 Ibid.

8 Ray D. Owen, 'Immunogenetic consequences of vascular anastomoses between bovine twins', *Science* 1945; 102(2651): 400–1.

9 Rupert E. Billingham, Leslie Brent and Peter B. Medawar, '"Actively acquired tolerance" of foreign cells', *Nature* 1953; 172: 603–6.

10 Medawar, *Memoir of a Thinking Radish*, 160.

11 Ibid., 134.

12 Brent, *A History of Transplantation Immunology*, 75.

13 Ibid.

14 G. H. Algire, J. M. Weaver and R. T. Prehn, 'Growth of cells in vivo in diffusion chambers. I. Survival of homografts in immunized mice', *Journal of the National Cancer Institute* 1954; 15(3): 493–507.

CHAPTER 19

1 F. R. Appelbaum, 'Hematopoietic-cell transplantation at 50', *New England Journal of Medicine* 2007; 357(15): 1472–5.
2 E. Donnall Thomas, 'Bone marrow transplantation – past, present and future', Nobel Lecture, 8 December 1990. Available at https://www.nobelprize.org/prizes/medicine/1990/thomas/lecture/.
3 Appelbaum, 'Hematopoietic-cell transplantation at 50'.
4 E. Donnall Thomas et al., 'Supralethal whole body irradiation and isologous marrow transplantation in man', *Journal of Clinical Investigation* 1959; 38(10): 1709–16.
5 Ibid.
6 Ibid.
7 Donnall Thomas, 'Bone marrow transplantation'.
8 Joseph Murray, 'The first successful organ transplants in man', Nobel Lecture, 8 December 1990. Available at https://www.nobelprize.org/prizes/medicine/1990/murray/lecture/.
9 Quoted in Elizabeth Simpson, 'Medawar's legacy to cellular immunology and clinical transplantation: a commentary on Billingham, Brent and Medawar (1956) "Quantitative studies on tissue transplantation immunity. III. Actively acquired tolerance"', *Philosophical Transactions of the Royal Society B: Biological Sciences* 2015; 370(1666): 20140382.
10 Murray, 'The first successful organ transplants in man'.
11 Robert Schwartz and William Dameshek, 'Drug-induced immunological tolerance', *Nature* 1959; 183(4676): 1682–3.
12 R. Y. Calne, 'Inhibition of the rejection of renal homografts in dogs by drugs', *Annals of the Royal College of Surgeons of England* 1963; 32(5): 281–302.
13 Simpson, 'Medawar's legacy'.

CHAPTER 20

1 Quoted in E. E. Vella, 'Belsen: medical aspects of a World War II concentration camp', *Journal of the Royal Army Medical Corps* 1984; 130: 34–59.
2 James L. Gowans, 'The lymphocyte – a disgraceful gap in medical knowledge', *Immunology Today* 1996; 17(6): 288–91.
3 Ibid.
4 Ivan M. Roitt, *Essential Immunology* (3rd edn, Oxford: Blackwell, 1977), 48.
5 Ibid.
6 Gowans, 'The lymphocyte'.
7 Ibid.
8 Vella, 'Belsen'.

CHAPTER 21

1   Ogden C. Bruton, 'Agammaglobulinemia', *Pediatrics* 1952; 9(6): 722–8; William R. Clark, *At War Within: The Double-Edged Sword of Immunity* (New York and Oxford: Oxford University Press, 1995), 62.

2   Eduard Glanzmann and Paul Riniker, 'Essential lymphocytophthisis; new clinical aspect of infant pathology', *Annales Paediatrici: International Review of Pediatrics* 1950; 175(1–2): 1–32.

3   Steve McVicker, 'Bursting the bubble', *Houston Press* (10 April 1997). Available at https://www.houstonpress.com/news/bursting-the-bubble-6573830; William R. Clark, *At War Within: The Double-Edged Sword of Immunity* (New York and Oxford: Oxford University Press, 1995), 71.

CHAPTER 22

1   Quoted in Dik Evans, 'Postsplenectomy sepsis 10 years or more after operation', *Journal of Clinical Pathology* 1985; 38: 309–31.

2   Bruce Glick, Timothy S. Chang and R. George Jaap, 'The bursa of Fabricius and antibody production', *Poultry Science* 1955; 35(1): 224–5.

3   Geoff Watts, 'Jacques Miller: immunologist who discovered role of the thymus', *Lancet* 2011; 378(9799): 1290.

4   Jacques F. A. P. Miller, 'Effect of neonatal thymectomy on the immunological responsiveness of the mouse', *Proceedings of the Royal Society of London* 1962; 156(964): 415–28.

5   Alexander D. Gitlin and Michel C. Nussenzweig, 'Immunology: fifty years of B lymphocytes', *Nature* 2015; 517(7533): 139–41.

6   Max D. Cooper, Raymond D. A. Peterson and Robert A. Good, 'Delineation of the thymic and bursal lymphoid systems in the chicken', *Nature* 1965; 205: 143–6.

7   Gitlin and Nussenzweig, 'Immunology: fifty years of B lymphocytes'.

8   Henry N. Claman, Edward A. Chaperon and R. Faser, 'Thymus-marrow cell combinations: synergism in antibody production', *Proceedings of the Society for Experimental Biology and Medicine* 1966; 122(4): 1167–71.

9   Jacques F. A. P. Miller, 'The Croonian lecture, 1992: the key role of the thymus in the body's defence strategies', *Philosophical Transactions of the Royal Society B: Biological Sciences* 1992; 337(1279): 105–24.

10 Reported in a comment on Max D. Cooper, Raymond D. A. Peterson and Robert A. Good, 'A new concept of the cellular basis of immunity', *Journal of Pediatrics* 1965; 67(5): 907.

CHAPTER 23

1 G. Norkrans et al., 'Toxic shock and tampons', *British Medical Journal* 1980; 281(6252): 1426.

2 Charles A. Dinarello, 'The history of fever, leukocytic pyrogen and interleukin-1', *Temperature (Austin)* 2015; 2(1): 8–16.

3 Igal Gery, 'The definition of lymphocyte activating factor: giving a helping hand to serendipity', *Frontiers in Immunology* 2014; 5: 610.

4 Barry R. Bloom et al., 'Demonstration of delayed hypersensitivity to soluble antigens of chemically induced tumors by inhibition of macrophage migration', *Proceedings of the National Academy of Sciences of the USA* 1969; 64(4): 1176–80.

5 Vincent C. Tam and Alan Aderem, 'Macrophage activation as an effector mechanism for cell-mediated immunity', *Journal of Immunology* 2014; 193(7): 3183–4.

6 George B. Mackaness, 'The influence of immunologically committed lymphoid cells on macrophage activity in vivo', *Journal of Experimental Medicine* 1969; 129(5): 973–92.

7 George B. Mackaness, 'Cellular resistance to infection', *Journal of Experimental Medicine* 1962; 116: 381–406.

8 George B. Mackaness and William C. Hill, 'The effect of anti-lymphocyte globulin on cell-mediated resistance to infection', *Journal of Experimental Medicine* 1969; 129(5): 993–1012.

9 George B. Mackaness, 'The immunological basis of acquired cellular resistance', *Journal of Experimental Medicine* 1964; 120(1): 105–20.

CHAPTER 24

1 Albert H. Coons, Elizabeth H. Leduc and Jeanne M. Connolly, 'Studies on antibody production. I. A method for the histochemical demonstration of specific antibody and its application to a study of the hyperimmune rabbit', *Journal of Experimental Medicine* 1955; 102(1): 49–60.

2 Linus Pauling, 'A theory of the structure and process of formation of antibodies', *Journal of the American Chemical Society* 1940; 62(10), 2643–57.

3 Niels K. Jerne, 'The natural-selection theory of antibody formation', *Proceedings of the National Academy of Sciences of the USA* 1955; 41(11): 849–57.

4 Domenico Ribatti, 'Sir Frank Macfarlane Burnet and the clonal selection theory of antibody formation', *Clinical and Experimental Medicine* 2009; 9(4): 253–8.

5 Gustav J. V. Nossal and Joshua Lederberg, 'Antibody production by single cells', *Nature* 1958; 181(4620): 1419–20.

6 'Scientists track "memory" B cells – responsible for lifelong immunity – to the bone marrow', Walter & Eliza Hall Institute of Medical Research [website]. Available at https://discovery.wehi.edu.au/timeline/immune-memory.

7   Ibid.
8   Ibid.

CHAPTER 25

1   Arne Tiselius and Elvin A. Kabat, 'An electrophoretic study of immune sera and purified antibody preparations', *Journal of Experimental Medicine* 1939; 69(1): 119–31.
2   Rodney R. Porter, 'The formation of a specific inhibitor by hydrolysis of rabbit antiovalbumin', *Biochemical Journal* 1950; 46(4): 479–87.
3   Rodney R. Porter, 'Chemical structure of gamma-globulin and antibodies', *British Medical Bulletin* 1963; 19(3): 197–201.
4   Robin C. Valentine and N. Michael Green, 'Electron microscopy of an antibody-hapten complex', *Journal of Molecular Biology* 1967; 27(3): 615–17.
5   Gerald M. Edelman and Joseph A. Gally, 'The nature of Bence-Jones proteins', *Journal of Experimental Medicine* 1962; 116(2): 207–27.
6   Ivan M. Roitt, *Essential Immunology* (3rd edn, Oxford: Blackwell, 1977), 29–30.
7   Gerald M. Edelman et al., 'The covalent structure of an entire gammaG immunoglobulin molecule', *Proceedings of the National Academy of Sciences of the USA* 1969; 63(1): 78–85.
8   Tae Te Wu and Elvin A. Kabat, 'An analysis of the sequences of the variable regions of Bence Jones proteins and myeloma light chains and their implications for antibody complementarity', *Journal of Experimental Medicine* 1970; 132(2): 211–50.
9   Linus Pauling, 'A theory of the structure and process of formation of antibodies', *Journal of the American Chemical Society* 1940; 62(10): 2643–57.
10 William J. Dreyer and J. Claude Bennett, 'The molecular basis of antibody formation: a paradox', *Proceedings of the National Academy of Sciences of the USA* 1965; 54(3): 864–9.
11 Susumu Tonegawa, 'Biographical', Nobel Prize [website]. Available at https://www.nobelprize.org/prizes/medicine/1987/tonegawa/biographical.
12 Ibid.
13 Susumu Tonegawa, 'Somatic generation of immune diversity', Nobel Lecture, 8 December 1987. Available at https://www.nobelprize.org/prizes/medicine/1987/tonegawa/lecture/.

CHAPTER 26

1   César Milstein, 'From the structure of antibodies to the diversification of the immune response', Nobel Lecture, 8 December 1984. Available at https://www.nobelprize.org/prizes/medicine/1984/milstein/lecture/.

2   Georges Kohler and César Milstein, 'Continuous cultures of fused cells secret-
    ing antibody of predefined specificity', *Nature* 1975; 256(5517): 495–7.
3   Lara Marks, 'The story of César Milstein and monoclonal antibod-
    ies', What Is Biotechnology? [website]. Available at http://www.whatis-
    biotechnology.org/index.php/exhibitions/milstein/introduction/
    Introduction-to-the-story-of-Cesar-Milstein-and-mAbs.
4   Lois A. Lampson and Ronald Levy, 'A role for clonal antigens in cancer diagnosis
    and therapy', *Journal of the National Cancer Institute* 1979; 62(2): 217–20.
5   Richard A. Miller et al., 'Treatment of B-cell lymphoma with monoclonal
    anti-idiotype antibody', *New England Journal of Medicine* 1982; 306: 517–22.
6   Josh Halliday and agencies, 'Cancer trial of drug combination yields "spectacu-
    lar" results', *Guardian* (1 June 2015). Available at https://www.theguardian.
    com/society/2015/jun/01/paradigm-shift-hailed-in-treatment-of-lung-cancer.

CHAPTER 27

1   Susan Okie, 'An AIDS clue in Kenya?', *Washington Post* (14 December 1993). Available
    at https://www.washingtonpost.com/archive/lifestyle/wellness/1993/12/14/
    an-aids-clue-in-kenya/4f6ca387-49c7-4880-b53c-e499f16966a7/.
2   Laurent Degos, 'Jean Dausset a scientific pioneer: intuition and creativity for
    the patients (1916–2009)', *Haematologica* 2009; 94(9): 1331–2.
3   Peter Gorer, 'The genetic and antigenic basis of tumour transplantation',
    Journal of Pathology and Bacteriology 1937; 44(3): 691–7.
4   George D. Snell, 'Methods for the study of histocompatibility genes', *Journal
    of Genetics* 1948; 49(2): 87–108.
5   Baruj Benacerraf, 'Biographical', Nobel Prize [website]. Available at https://
    www.nobelprize.org/prizes/medicine/1980/benacerraf/biographical.
6   Ibid.
7   Baruj Benacerraf, 'The role of MHC gene products in immune regulation and
    its relevance to alloreactivity', Nobel Lecture, 8 December 1990. Available at
    https://www.nobelprize.org/prizes/medicine/1980/benacerraf/lecture/.
8   Ibid.
9   E. R. Unanue and J. C. Cerottini, 'The function of macrophages in the immune
    response', *Seminars in Hematology* 1970; 7(2): 225–48.
10 Emil R. Unanue, 'The regulatory role of macrophages in antigenic stimula-
    tion', *Advances in Immunology* 1972; 15: 95–165.
11 Bruce P. Babbitt et al., 'Binding of immunogenic peptides to Ia histocompat-
    ibility molecules', *Nature* 1985; 317(6035): 359–61.
12 Benacerraf, 'The role of MHC gene products'.
13 Jean Dausset, 'The major histocompatibility complex in man: past, present,
    and future concepts', Nobel Lecture, 8 December 1980. Available at https://
    www.nobelprize.org/prizes/medicine/1980/dausset/lecture/.

14 Rolf M. Zinkernagel and Peter C. Doherty, 'Restriction of *in vitro* T cell-mediated cytotoxicity in lymphocytic choriomeningitis within a syngeneic or semiallogeneic system', *Nature* 1974; 248: 701–2.

15 Roger Beckman, 'Professor Peter Doherty, immunologist' [interview], Australian Academy of Science [website]. Available at https://www.science.org.au/learning/general-audience/history/interviews-australian-scientists/professor-peter-doherty#2.

16 Jenefer M. Blackwell, Sarra E. Jamieson and David Burgner, 'HLA and infectious diseases', *Clinical Microbiology Reviews* 2009; 22(2): 370–85.

CHAPTER 28

1 Gina Kolata, 'Boy's 1969 death suggests AIDS invaded U.S. several times', *New York Times* (28 October 1987). Available at https://www.nytimes.com/1987/10/28/us/boy-s-1969-death-suggests-aids-invaded-us-several-times.html.

2 UNAIDS Fact Sheet 2016.

3 E. A. Engels et al., 'Trends in cancer risk among people with AIDS in the United States 1980–2002', *AIDS* 2006; 20(12): 1645–54.

4 Dejan Pavlovic et al., 'Progressive multifocal leukoencephalopathy: current treatment options and future perspectives', *Therapeutic Advances in Neurological Disorders* 2015; 8(6): 255–73.

5 A. G. Abraham at al., 'Invasive cervical cancer risk among HIV-infected women: a North American multicohort collaboration prospective study', *Journal of Acquired Immune Deficiency Syndromes* 2013; 62(4): 405–13.

CHAPTER 29

1 D. H. Katz and E. R. Unanue, 'Critical role of determinant presentation in the induction of specific responses in immunocompetent lymphocytes', *Journal of Experimental Medicine* 1973; 137(4): 967–90.

2 Daniel Engber, 'Is the cure for cancer inside you?', *New York Times* (21 December 2012). Available at http://www.nytimes.com/2012/12/23/magazine/is-the-cure-for-cancer-inside-you.html.

3 Ralph M. Steinman (delivered by Michel C. Nussenzweig), 'Ralph Steinman and the discovery of dendritic cells', Nobel Lecture, 7 December 2011. Available at https://www.nobelprize.org/prizes/medicine/2011/steinman/lecture/.

4 Ibid.

5 Ibid.

6 Ibid.

7 Ibid.

8   Ibid.

9   Ibid.

10 Ibid.

11 Engber, 'Is the cure for cancer inside you?'

12 Quoted in Lauren Gravitz, 'A fight for life that united a field', *Nature* 2011; 478(7368): 163–4.

13 Quoted ibid.

14 Gravitz, 'A fight for life that united a field'.

15 Quoted ibid.

16 Quoted in Julie Steenhuysen and Michelle Nichols, 'Insight: Nobel winner's last big experiment: himself', Reuters [website] (6 October 2011). Available at https://www.reuters.com/article/us-nobel-medicine-experiment/insight-nobel-winners-last-big-experiment-himself-idUSTRE7956CN20111006.

17 'Ralph Steinman remains Nobel Laureate' [press release], Nobel Foundation [website] (3 October 2011). Available at http://news.cision.com/nobelstiftelsen/r/ralph-steinman-remains-nobel-laureate,c9169294.

18 Steinman, 'Ralph Steinman and the discovery of dendritic cells'.

CHAPTER 30

1   Bin Zhang et al., 'Treatment with convalescent plasma for critically ill patients with severe acute respiratory syndrome coronavirus 2 infection', *Chest* 2020; 158(1): e9–e13.

2   P. Bégin et al., 'Convalescent plasma for hospitalized patients with COVID-19: an open-label, randomized controlled trial', *Nature Medicine* 2021; 28(1): 212.

3   Giuseppe Marano et al., 'Convalescent plasma: new evidence for an old therapeutic tool?', *Blood Transfusion* 2016; 14(2): 152–7.

4   Thomas C. Luke et al., 'Meta-analysis: convalescent blood products for Spanish influenza pneumonia: a future H5N1 treatment?', *Annals of Internal Medicine* 2006; 145(8): 599–609.

5   Kuo-Ming Yeh et al., 'Experience of using convalescent plasma for severe acute respiratory syndrome among healthcare workers in a Taiwan hospital', *Journal of Antimicrobial Chemotherapy* 2005; 56(5): 919–22.

6   Linlin Bao et al., 'Reinfection could not occur in SARS-CoV-2 infected rhesus macaques' [preprint], bioRxiv [website] (2020). Available at https://www.biorxiv.org/content/10.1101/2020.03.13.990226v1.

7   Irani Thevarajan et al., 'Breadth of concomitant immune responses prior to patient recovery: a case report of non-severe COVID-19', *Nature Medicine* 2020; 26(4): 453–5.

8   Marano et al., 'Convalescent plasma'.

9   Zhiqiang Ku et al., 'Antibody therapies for the treatment of COVID-19', *Antibody Therapeutics* 2020; 3(2): 101–8.

10 G. Salazar et al., 'Antibody therapies for the prevention and treatment of viral infections', *NPJ Vaccines* 2017; 2: 19.

11 Bin Ju et al., 'Potent human neutralizing antibodies elicited by SARS-CoV-2 infection', *Nature* 2020; 584: 115–19.

12 Ku et al., 'Antibody therapies for the treatment of COVID-19'.

13 M. S. Diamond and T. C. Pierson, 'Molecular insight into dengue virus pathogenesis and its implications for disease control', *Cell* 2015; 162(3): 488–92.

14 Li Liu et al., 'Anti-spike IgG causes severe acute lung injury by skewing macrophage responses during acute SARS-CoV infection', *JCI Insight* 2019; 4(4): e123158.

15 Qing Ye et al., 'The pathogenesis and treatment of the "cytokine storm" in Covid-19', *Journal of Infection* 2020; 80(6): 607–13.

CHAPTER 31

1 Quoted in Greg Myre, 'Among the 1st to get a polio vaccine, Peter Salk says don't rush a COVID-19 shot', NPR [website] (30 May 2020). Available at https://www.npr.org/2020/05/30/861887610/among-the-1st-to-get-a-polio-vaccine-peter-salk-says-dont-rush-a-covid-19-shot.

2 Jason Beaubien, 'Wiping out polio: how the U.S. snuffed out a killer', NPR [website] (15 October 2012). Available at https://www.npr.org/sections/health-shots/2012/10/16/162670836/wiping-out-polio-how-the-u-s-snuffed-out-a-killer.

3 Ibid.

4 Dennis Denenberg and Lorraine Roscoe, *50 American Heroes Every Kid Should Meet* (2nd edn, Minneapolis: Millbrook, 2016), 91.

5 Sarah Boden, 'Site where polio vaccine was first tested on humans to receive state historic marker', 90.5 WESA [website] (2 April 2018). Available at https://www.wesa.fm/post/site-where-polio-vaccine-was-first-tested-humans-receive-state-historic-marker.

6 Beaubien, 'Wiping out polio'.

7 Qiang Gao et al., 'Development of an inactivated vaccine candidate for SARS-CoV-2', *Science* 2020; 369(6499): 77–81.

8 Stanley Plotkin, 'History of vaccination', *Proceedings of the National Academy of Science of the USA* 2014; 111(34): 12283–7.

9 Ibid.

10 Michael Fitzpatrick, 'The Cutter incident: how America's first polio vaccine led to a growing vaccine crisis', *Journal of the Royal Society of Medicine* 2006; 99(3): 156.

11 Siang Yong Tan and Nate Ponstein, 'Jonas Salk (1914–1995): a vaccine against polio', *Singapore Medical Journal* 2019; 60(1): 9–10.

12 Neeltje van Doremalen et al., 'ChAdOx1 nCoV-19 vaccination prevents SARS-CoV-2 pneumonia in rhesus macaques', *Nature* 2020; 586: 578–82.

13 Neeltje van Doremalen et al., 'A single dose of ChAdOx1 MERS provides protective immunity in rhesus macaques', *Science Advances* 2020; 6(24): eaba8399.

14 Karuppiah Muthumani et al., 'A synthetic consensus anti-spike protein DNA vaccine induces protective immunity against Middle East respiratory syndrome coronavirus in nonhuman primates', *Science Translational Medicine* 2015; 7(301): 301ra132.

15 Trevor R. F. Smith et al., 'Immunogenicity of a DNA vaccine candidate for COVID-19', *Nature Communications* 2020; 11(1): 2601.

16 David Liebowitz et al., 'High titre neutralising antibodies to influenza after oral tablet immunisation: a phase 1, randomised, placebo-controlled trial', *Lancet Infectious Diseases* 2015; 15(9): 1041–48.

17 Quoted in Myre, 'Among the 1st to get a polio vaccine'.

CHAPTER 32

1 Office for National Statistics, UK reports, June 2020.

2 Tetsuro Araki et al., 'Normal thymus in adults: appearance on CT and associations with age, sex, BMI and smoking', *European Radiology* 2016; 26(1): 15–24.

3 Kengo Yoshida et al., 'Inverse associations between obesity indicators and thymic T-cell production levels in aging atomic-bomb survivors', *PLOS One* 2014; 9(3): e91985.

4 Soo-Jin Oh et al., 'Aging and the immune system: the impact of immunosenescence on viral infection, immunity and vaccine immunogenicity', *Immune Network* 2019; 19(6): e37.

5 Vasileios Kordinas et al., 'The telomere/telomerase system in chronic inflammatory diseases. Cause or effect?', *Genes (Basel)* 2016; 7(9): 60.

6 Jennifer B. Dowd et al., 'Persistent herpesvirus infections and telomere attrition over 3 years in the Whitehall II cohort', *Journal of Infectious Diseases* 2017; 216(5): 565–72.

7 Oh et al., 'Aging and the immune system'.

8 Michelle A. Murray and Sanjay H. Chotirmall, 'The impact of immunosenescence on pulmonary disease', *Mediators of Inflammation* 2015; 2015: 1–10.

9 Richard J. Simpson and Emmanuel Katsanis, 'The immunological case for staying active during the COVID-19 pandemic', *Brain, Behavior, and Immunity* 2020; 87: 6–7.

10 Stephen A. Martin, Brandt D. Pence and Jeffrey A. Woods, 'Exercise and respiratory tract viral infections', *Exercise and Sport Sciences Reviews* 2009; 37(4): 157–64.

11 R. Ader, 'Psychoneuroimmunology', in Neil J. Smelser and Paul B. Baltes, eds, *International Encyclopedia of the Social & Behavioral Sciences* (Oxford: Elsevier, 2001), xviii, 12422–8.

12 David S. Black and George M. Slavich, 'Mindfulness meditation and the immune system: a systematic review of randomized controlled trials', *Annals of the New York Academy of Sciences* 2016; 1373(1): 13–24.

13 J. E. Bower et al., 'Mindfulness meditation for younger breast cancer survivors: a randomized controlled trial', *Cancer* 2015; 121(8): 1231–40.

14 S. Jedel et al., 'A randomized controlled trial of mindfulness-based stress reduction to prevent flare-up in patients with inactive ulcerative colitis', *Digestion* 2014; 89(2): 142–55.

15 J. D. Creswell et al., 'Mindfulness-based stress reduction training reduces loneliness and pro-inflammatory gene expression in older adults: a small randomized controlled trial', *Brain, Behavior, and Immunity* 2012; 26(7): 1095–101.

16 Bower et al., 'Mindfulness meditation for younger breast cancer survivors'.

17 T. W. Pace, et al., 'Increased stress-induced inflammatory responses in male patients with major depression and increased early life stress', *American Journal of Psychiatry* 2006; 163(9): 1630–3.

18 J. D. Creswell, H.F. Myers, S.W. Cole and M.R. Irwin, 'Mindfulness meditation training effects on CD4+ T lymphocytes in HIV-1 infected adults: a small randomized controlled trial', *Brain, Behaviour, and Immunity* 2009; 23(2): 184–8.

19 Jennifer Daubenmier et al., 'Changes in stress, eating, and metabolic factors are related to changes in telomerase activity in a randomized mindfulness intervention pilot study', *Psychoneuroendocrinology* 2012; 37(7): 917–28.

20 Quoted in Zoë Corbyn, 'Elizabeth Blackburn on the telomere effect: "It's about keeping healthier for longer"', *Observer* (29 January 2017). Available at https://www.theguardian.com/science/2017/jan/29/telomere-effect-elizabeth-blackburn-nobel-prize-medicine-chromosomes.

Index